Penguin Books **Structures**

James Edward Gordon was born in 1913. He took a degree in naval architecture at Glasgow University and worked in wood and steel shipyards, intending to design sailing ships. On the outbreak of the Second World War he moved to the Royal Aircraft Establishment at Farnborough, where he worked on wooden aircraft, plastics and unorthodox materials of all kinds. He designed the sailing rescue dinghies carried at one time by most bomber aircraft. He later became head of the plastics structures section at Farnborough and developed a method of construction in asbestos-reinforced plastics which is now used for a number of purposes in aircraft and rockets. For several frustrating years he worked in industry on the strength of glass and the growth of strong 'whisker' crystals. In 1962 he returned to government service as superintendent of an experimental branch at Waltham Abbey concerned with research and development of entirely new structural materials – most of which are based on 'whiskers'. He was Industrial Fellow Commoner at Churchill College, Cambridge, and became Professor of Materials Technology at the University of Reading, where he is now Professor Emeritus. He has been awarded the British Silver Medal of the Royal Aeronautical Society for work on aircraft plastics and also the Griffith Medal of the Materials Science Club for contributions to materials science. His book *The New Science of Strong Materials* has also been published in Penguins.

Professor J. E. Gordon is married and has two grown-up sons. His spare-time interests are sailing, Greek, photography and ski-ing.

J. E. Gordon **Structures**

or Why Things Don't Fall Down

Penguin Books

PENGUIN BOOKS

Published by the Penguin Group
Penguin Books Ltd, 27 Wrights Lane, London W8 5TZ, England
Penguin Books USA Inc., 375 Hudson Street, New York, New York 10014, USA
Penguin Books Australia Ltd, Ringwood, Victoria, Australia
Penguin Books Canada Ltd, 10 Alcorn Avenue, Toronto, Ontario, Canada M4V 3B2
Penguin Books (NZ) Ltd, 182–190 Wairau Road, Auckland 10, New Zealand

Penguin Books Ltd, Registered Offices: Harmondsworth, Middlesex, England

First published in Pelican Books 1978
Reprinted in Penguin Books 1991
10 9 8 7 6 5 4 3 2 1

Printed in England by Clays Ltd, St Ives plc
Set in Monotype Times

To my grandchildren,
Timothy and Alexander

Among the innumerable mortifications which waylay human arrogance on every side may well be reckoned our ignorance of the most common objects and effects, a defect of which we become more sensible by every attempt to supply it. Vulgar and inactive minds confound familiarity with knowledge and conceive themselves informed of the whole nature of things when they are shown their form or told their use; but the speculatist, who is not content with superficial views, harasses himself with fruitless curiosity, and still, as he inquires more, perceives only that he knows less.

Samuel Johnson, *The Idler* (Saturday, 25 November 1758)

Contents

List of Plates

Foreword

I am very much aware that it is an act of extreme rashness to attempt to write an elementary book about structures. Indeed it is only when the subject is stripped of its mathematics that one begins to realize how difficult it is to pin down and describe those structural concepts which are often called 'elementary'; by which I suppose we mean 'basic' or 'fundamental'. Some of the omissions and oversimplifications are intentional but no doubt some of them are due to my own brute ignorance and lack of understanding of the subject.

Although this volume is more or less a sequel to *The New Science of Strong Materials* it can be read as an entirely separate book in its own right. For this reason a certain amount of repetition has been unavoidable in the earlier chapters.

I have to thank a great many people for factual information, suggestions and for stimulating and sometimes heated discussions. Among the living, my colleagues at Reading University have been generous with help, notably Professor W. D. Biggs (Professor of Building Technology), Dr Richard Chaplin, Dr Giorgio Jeronimidis, Dr Julian Vincent and Dr Henry Blyth; Professor Anthony Flew, Professor of Philosophy, made useful suggestions about the last chapter. I am also grateful to Mr John Bartlett, Consultant Neurosurgeon at the Brook Hospital. Professor T. P. Hughes of the University of the West Indies has been helpful about rockets and many other things besides. My secretary, Mrs Jean Collins, was a great help in times of trouble. Mrs Nethercot of *Vogue* was kind to me about dressmaking. Mr Gerald Leach and also many of the editorial staff of Penguins have exercised their accustomed patience and helpfulness.

Among the dead, I owe a great deal to Dr Mark Pryor – lately of Trinity College, Cambridge – especially for discussions about biomechanics which extended over a period of nearly thirty years. Lastly, for reasons which must surely be obvious, I owe a humble oblation to Herodotus, once a citizen of Halicarnassus.

Acknowledgements

We acknowledge with gratitude permission to quote from various authors. For Douglas English's poem, Punch Publications Ltd; for quotations from Weston Martyr's *The Southseaman*, Messrs. William Blackwood Ltd; for the quotation from Rudyard Kipling's *The Ship that Found Herself*, Messrs. A. P. Watt & Son and the executors of the late Mrs Bambridge and the Macmillan Co. of London and Basingstoke. Also to Mr H. L. Cox for the quotation from his book *The Design of Structures of Least Weight*. The quotations from the New English Bible (Second Edition © 1970) are by kind permission of the Oxford and Cambridge University Presses.

We are also most grateful to all those named in the List of Plates who have so kindly provided illustrations and given permission to reproduce them.

We have received a great deal of help from many people and organizations with regard to both quotations and illustrations. If we have, in any instance, failed to make proper acknowledgement, we offer our apologies.

Chapter 1   The structures in our lives

– or how to communicate with engineers

> *As men journeyed in the east, they came upon a plain*
> *in the land of Shinar and settled there. They said to*
> *one another, 'Come, let us make bricks and bake*
> *them hard'; they used bricks for stones and bitumen*
> *for mortar. 'Come,' they said, 'let us build ourselves*
> *a city and a tower with its top in the heavens, and*
> *make a name for ourselves; or we shall be dispersed*
> *all over the earth.' Then the Lord came down to see*
> *the city and tower which mortal men had built, and*
> *he said, 'Here they are, one people with a single*
> *language, and now they have started to do this;*
> *henceforward nothing they have a mind to do will be*
> *beyond their reach. Come, let us go down there and*
> *confuse their speech, so that they will not understand*
> *what they say to one another.' So the Lord dispersed*
> *them from there all over the earth, and they left off*
> *building the city. That is why it is called Babel (that*
> *is, Babylon), because the Lord there made a babble*
> *of the language of all the world.*

> *Genesis 11. 2–9 (New English Bible)*

A structure has been defined as 'any assemblage of materials
which is intended to sustain loads', and the study of structures is
one of the traditional branches of science. If an engineering structure breaks, people are likely to get killed, and so engineers do
well to investigate the behaviour of structures with circumspec-
tion. But, unfortunately, when they come to tell other people
about their subject, something goes badly wrong, for they talk in
a strange language, and some of us are left with the conviction
that the study of structures and the way in which they carry loads
is incomprehensible, irrelevant and very boring indeed.

Yet structures are involved in our lives in so many ways that we
cannot really afford to ignore them: after all, every plant and
animal and nearly all of the works of man have to sustain greater

or less mechanical forces without breaking, and so practically everything is a structure of one kind or another. When we talk about structures we shall have to ask, not only why buildings and bridges fall down and why machinery and aeroplanes sometimes break, but also how worms came to be the shape they are and why a bat can fly into a rose-bush without tearing its wings. How do our tendons work? Why do we get 'lumbago'? How were pterodactyls able to weigh so little? Why do birds have feathers? How do our arteries work? What can we do for crippled children? Why are sailing ships rigged in the way they are? Why did the bow of Odysseus have to be so hard to string? Why did the ancients take the wheels off their chariots at night? How did a Greek catapult work? Why is a reed shaken by the wind and why is the Parthenon so beautiful? Can engineers learn from natural structures? What can doctors and biologists and artists and archaeologists learn from engineers?

As it has turned out, the struggle to understand the real reasons why structures work and why things break has been a great deal more difficult and has taken much longer than one might have expected. It is really only quite recently that we have been able to fill in enough of the gaps in our knowledge to answer some of these questions in any very useful or intelligent manner. Naturally, as more of the bits of the jig-saw puzzle are assembled, the general picture becomes clearer: the whole subject is becoming less a study for rather narrow specialists and more one which the ordinary person can find rewarding and relevant to a wide range of general interests.

This book is about modern views on the structural element in Nature, in technology and in everyday life. We shall discuss the ways in which the need to be strong and to support various necessary loads has influenced the development of all sorts of creatures and devices – including man.

The living structure

Biological structures came into being long before artificial ones. Before there was life in the world, there was no such thing as a purposive structure of any kind – only mountains and heaps of sand and rock. Even a very simple and primitive kind of life is a delicately balanced, self-perpetuating chemical reaction which needs to be separated and guarded from non-life. Nature having invented life – and with it individualism – it became necessary to devise some kind of container in which to keep it. This film or membrane had to have at least a minimum of mechanical strength, both to contain the living matter and also to give it some protection from outside forces.

If, as seems possible, some of the earliest forms of life consisted of tiny droplets floating in water, then a very weak and simple barrier, perhaps no more than the surface tension which exists at the interfaces between different liquids, may have sufficed. Gradually, as living creatures multiplied, life became more competitive, and the weak, globular and immobile animals were at a disadvantage. Skins became tougher and various means of locomotion were evolved. Larger, multicellular animals appeared which could bite and could swim fast. Survival became a matter of chasing and being chased, eating and being eaten. Aristotle called this *allelophagia* – a mutual eating – Darwin called it natural selection. In any case, progress in evolution was dependent upon the development of stronger biological materials and more ingenious living structures.

The earlier and more primitive animals were mostly made from soft materials because they not only make it much easier to wriggle and extend oneself in various ways, but soft tissues are usually tough (as we shall see), while rigid ones like bone are often brittle. Furthermore, the use of rigid materials imposes all kinds of difficulties in connection with growth and reproduction. As women know, the business of giving birth involves an engineering of high strains and large deflections. All the same, the development of the vertebrate foetus from conception onwards, like that of natural structures in general, is in certain respects from soft to hard, and the hardening process goes on after the baby has emerged.

One gets the impression that Nature has accepted the use of stiff materials rather reluctantly, but, as animals got bigger and came out of the water on to the land, most of them developed and exploited rigid skeletons, teeth and sometimes horns and armour. Yet animals never became predominantly rigid devices like most modern machinery. The skeleton usually remained but a small part of the whole, and, as we shall see, the soft parts were frequently used in clever ways to limit the loads upon the skeleton and thus to protect it from the consequences of its brittleness.

While the bodies of most animals are made preponderantly from flexible materials, this is not always true for plants. The smaller and more primitive plants are usually soft, but a plant cannot chase its food, nor can it run away from an enemy. It can, however, protect itself to some extent by growing tall, and, by doing so, it may also be able to get more than its fair share of sun and rain. Trees, in particular, seem to be extraordinarily clever at stretching out to collect the diffuse and fitful energy of sunlight and at the same time standing up to being bullied by the wind – and all in the most cost-effective way. The tallest trees reach a height of about 360 feet or 110 metres, being by far the largest and most durable of living structures. For a plant to reach even a tenth of this height, however, its main structure needs to be both light and rigid; we shall see later that it incorporates a number of important lessons for engineers.

It may seem obvious that questions like these about strength and flexibility and toughness are relevant in medicine and in zoology and botany, yet for a long time both doctors and biologists resisted all such ideas with considerable success and with the whole force of their emotions. Of course, it is partly a matter of temperament and partly a matter of language, and perhaps a dislike and fear of the mathematical concepts of the engineer may have had something to do with the matter. Too often biologists simply cannot bring themselves to make a sufficiently serious study of the structural aspects of their problems. Yet there can be no reason to assume that, while Nature uses methods of infinite subtlety in her chemistry and her control mechanisms, her structural approach should be a crude one.

The technological structure

> *Wonders there are many, but there is no wonder*
> *Wilder than man –*
> *Man who makes the winds of winter bear him,*

Through the trough of waves that tower about him,
Across grey wastes of sea;
Man who wearies the Untiring, the Immortal –
Earth, eldest of the Gods, as year by year,
His plough teams come and go.
The care-free bands of birds,
Beasts of the wild, tribes of the sea,
In netted toils he takes,
The Subtle One.

Sophocles, *Antigone* (440 B.C.; translated by F. L. Lucas)

Benjamin Franklin (1706–90) used to define man as 'a tool-making animal'. In fact a good many other animals make and use rather primitive tools, and of course they quite often make better houses than do many uncivilized men. It might not be very easy to point out the exact moment in the development of man at which his technology could be said noticeably to surpass that of the beasts that perish. Perhaps it was later than we think, especially if the early men were arboreal.

However this may be, the gap both in time and in technical achievement between the sticks and stones of the earliest men – which were not much better than the tools used by the higher animals – and the sophisticated and beautiful artefacts of the late Stone Age is an immense one. Pre-metallic cultures have survived in remote places until only yesterday and many of their devices can be seen and admired in museums. To make strong structures without the benefit of metals requires an instinct for the distribution and direction of stresses which is by no means always possessed by modern engineers; for the use of metals, which are so conveniently tough and uniform, has taken some of the intuition and also some of the thinking out of engineering. Since the invention of Fibreglass and other artificial composite materials we have been returning at times to the sort of fibrous non-metallic structures which were developed by the Polynesians and the Eskimoes. As a result we have become more aware of our own inadequacies in visualizing stress systems and, just possibly, more respectful of primitive technologies.

As a matter of fact the introduction of the technological metals to the civilized world – probably between 2,000 and 1,000 B.C. –

did not make a very large or immediate difference to most artificial structures, because metals were scarce, expensive and not very easy to shape. The use of metals for cutting tools and weapons and, to some extent, for armour had its effect, but the majority of load-bearing artefacts continued to be made from masonry and from timber and leather and rope and textiles.

Using the old mixed constructions, the millwright and the coachbuilder, the shipwright and the rigger, needed a very high degree of skill, though of course they had their blind spots and they made the sort of mistakes one might expect from men without a formal analytical training. On the whole, the introduction of steam and machinery resulted in a dilution of skills, and it also limited the range of materials in general use in 'advanced technology' to a few standardized, rigid substances such as steel and concrete.

The pressures in some of the early engines were not much higher than our blood-pressure but, since materials like leather are incapable of withstanding hot steam, the engineer could not contrive a steam engine out of bladders and membranes and flexible tubes. So he was compelled to evolve from metals, by mechanical means, movements which an animal might have achieved more simply and perhaps with less weight.* He had to get his effects by means of wheels, springs, connecting rods and pistons sliding in cylinders.

Although these rather clumsy devices were originally imposed on him by the limitations of his materials, the engineer has come to look on this kind of approach to technology as the only proper and respectable one. Once he has settled in his rut of metal cogwheels and girders the engineer takes a lot of shifting. Moreover this attitude to materials and technology has rubbed off on the general public. Not long ago, at a cocktail party, the pretty wife of an American scientist said to me 'So you're really telling me that people used to make *airplanes* out of *wood*? – out of *lumber*! I don't believe you, you're kidding me.'

To what extent this outlook is objectively justified and how far it is based on prejudice and a morbid passion for being up to date is one of the questions which we shall discuss in this book. We

* Compare pistons and bellows.

need to take a balanced view. The traditional range of engineering structures made from bricks and stone and concrete and from steel and aluminium have been very successful, and clearly we ought to take them seriously, both for their own sakes and for what they have to teach us in a broader context. We might remember, however, that the pneumatic tyre, for instance, has changed the face of land transport and is probably a more important invention than the internal combustion engine. Yet we do not often teach engineering students about tyres, and there has been a distinct tendency in the schools of engineering to sweep the whole business of flexible structures under the carpet. When we come to look at the question in a broad way we may perhaps find that, for solid quantitative reasons, there is a case for trying to rebuild some part of traditional engineering upon models which may well turn out to be partly biological in inspiration.

Whatever view we may take of these matters we cannot get away from the fact that every branch of technology must be concerned, to a greater or less extent, with questions of strength and deflections; and we may consider ourselves lucky if our mistakes in these directions are merely annoying or expensive and do not kill or injure somebody. Those concerned with electrical affairs might be reminded that a great proportion of the failures in electrical and electronic devices are mechanical in origin.

Structures can, and do, break, and this may be important and sometimes dramatic; but, in conventional technology, the rigidity and deflections of a structure before it breaks are likely to be more important in practice. A house, a floor or a table which wobbled or swayed would not be acceptable, and we should consider that the performance of, say, an optical device such as a microscope or a camera depends not only upon the quality of its lenses but also upon the accuracy and rigidity with which they are positioned. Faults of this kind are far too common.

Structures and aesthetics

> Could I find a place to be alone with heaven,
>   I would speak my heart out: heaven is my need.
> Every woodland tree is flushing like the dogwood,
>   Flashing like the whitebeam, swaying like the reed.

Flushing like the dogwood crimson in October;
Streaming like the flag-reed South-West blown;
Flashing as in gusts the sudden-lighted whitebeam:
All seem to know what is for heaven alone.

George Meredith, *Love in the Valley*

Nowadays, whether we like it or not, we are stuck with one form or another of advanced technology and we have got to make it work safely and efficiently: this involves, among other things, the intelligent application of structural theory. However, man does not live by safety and efficiency alone, and we have to face the fact that, visually, the world is becoming an increasingly depressing place. It is not, perhaps, so much the occurrence of what might be described as 'active ugliness' as the prevalence of the dull and the commonplace. Far too seldom is the heart rejoiced or does one feel any better or happier for looking at the works of modern man.

Yet most of the artefacts of the eighteenth century, even quite humble and trivial ones, seem to many of us to be at least pleasing and sometimes incomparably beautiful. To that extent people – all people – in the eighteenth century lived richer lives than most of us do today. This is reflected in the prices we pay nowadays for period houses and antiques. A society which was more creative and self-confident would not feel quite so strong a nostalgia for its great-grandfathers' buildings and household goods.

Although such a book as this is not the place in which to develop elaborate and perhaps controversial theories of applied art, the question cannot be wholly ignored. As we have said, nearly every artefact is in some sense a structure of one kind or another, and, although most artefacts are not primarily concerned with making an emotional or aesthetic effect, it is highly important to realize that there can be no such thing as an emotionally neutral statement. This is true whether the medium be speech or writing or painting or technological design. Whether we mean it or not, every single thing we design and make will have some kind of subjective impact, for good or bad, over and above its overt rational purpose.

I think we are up against yet another problem of communication. Most engineers have had no aesthetic training at all, and the tendency in the schools of engineering is to despise such matters as frivolous. In any case, there is little enough time in the crowded syllabuses. Modern architects have made it very clear to me that they cannot spare time from their lordly sociological objectives to consider such minor matters as the strength of their buildings; nor, indeed, can they spare much time for aesthetics, in which their clients are probably not much interested anyway. Again, furniture designers, incredibly, are not taught during their formal training how to calculate the deflection in an ordinary bookshelf when it is loaded with books, and so it is not very surprising that most of them seem to have no ideas about relating the appearance to the structure of their products.

The theory of elasticity, or why things do fall down

Or those eighteen, upon whom the tower in Siloam fell, and slew them, think you that they were sinners above all men that dwelt in Jerusalem?

Luke 13.4

Many people – especially English people – dislike theory, and usually they do not think very much of theoreticians. This seems to apply especially to questions of strength and elasticity. A really surprising number of people who would not venture into the fields of, say, chemistry or medicine feel themselves competent to produce a structure upon which someone's life may depend. If pressed, they might admit that a large bridge or an aeroplane was a little beyond them, but the common structures of life surely present only the most trivial of problems?

This is not to suggest that the construction of an ordinary shed is a matter calling for years of study; yet it is true that the whole subject is littered with traps for the unwary, and many things are not as simple as they might seem. Too often the engineers are only called in, professionally, to deal with the structural achievements of 'practical' men at the same time as the lawyers and the undertakers.

Nevertheless, for long centuries the practical men managed after their own fashion – at least in certain fields of construction. If you go and look at a cathedral you may well wonder whether you are impressed more deeply by the skill or by the faith of the people who built it. These buildings are not only of very great size and height; some of them seem to transcend the dull and heavy nature of their constructional materials and to soar upwards into art and poetry.

On the face of it it would seem obvious that the medieval masons knew a great deal about how to build churches and cathedrals, and of course they were often highly successful and superbly good at it. However, if you had had the chance to ask the Master Mason how it was really done and why the thing stood up at all, I think he might have said something like 'The building is kept up by the hand of God – always provided that, when we built it, we duly followed the traditional rules and mysteries of our craft.'

Naturally, the buildings we see and admire are those which have survived: in spite of their 'mysteries' and their skill and experience, the medieval masons were by no means always successful. A fair proportion of their more ambitious efforts fell down soon after they were built, or sometimes during construction. However, these catastrophes were just as likely to be regarded as sent from Heaven, to punish the unrighteous or to bring sinners to repentance, as to be the consequence of mere technical ignorance – hence the need for the remark about the tower of Siloam.*

Perhaps because they were too much obsessed by the moral significance of good workmanship, the old builders and carpenters and shipwrights never seem to have thought at all, in any scientific sense, about why a structure is able to carry a load. Professor Jacques Heyman has shown conclusively that the cathedral masons, at any rate, did not think or design in the modern way. Although some of the achievements of the medieval craftsmen are impressive, the intellectual basis of their 'rules' and 'mysteries' was not very different from that of a cookery book. What these people did was to make something very much like what had been made before.

As we shall see in Chapter 9, masonry is a rather exceptional case and there are some special reasons why it is sometimes safe

*There is an interesting discussion of pagan views on this subject in Gilbert Murray's *Five Stages of Greek Religion* (O.U.P., 1930). Again, the whole question of animism in connection with structures is worthy of study.

and practicable to scale up from small churches to large cathedrals, relying simply on experience and traditional proportions. For other kinds of structures this way of doing things will not work and is quite unsafe. This is the reason why, though buildings got bigger and bigger, for a very long time the size of the largest ships remained virtually constant. So long as there was no scientific way of predicting the safety of technological structures, attempts to make devices which were new or radically different were only too likely to end in disaster.

Thus, for generation after generation, men turned their heads away from a rational approach to problems of strength. However, if you make a habit of shelving questions which, in your secret heart, you must surely know to be important, the psychological consequences will be unhappy. What happened was just what one would have expected. The whole subject became a breeding-ground for cruelty and superstition. When a ship is christened by some noble matron with a bottle of champagne, or when a foundation stone is laid by a fat mayor, these ceremonies are the last vestiges of certain very nasty sacrificial rites.

During the course of the Middle Ages the Church managed to suppress most of the sacrifices, but it did not do much to encourage any kind of scientific approach. To escape completely from such attitudes – or to accept that God may work through the agency of the laws of science – requires a complete change of thinking, a mental effort such as we can scarcely comprehend today. It called for a quite exceptional combination of imagination with intellectual discipline at a time when the very vocabulary of science barely existed.

As it turned out, the old craftsmen never accepted the challenge, and it is interesting to reflect that the effective beginnings of the serious study of structures may be said to be due to the persecution and obscurantism of the Inquisition. In 1633, Galileo (1564–1642) fell foul of the Church on account of his revolutionary astronomical discoveries, which were considered to threaten the very bases of religious and civil authority. He was most firmly headed off astronomy and, after his famous recantation,* he was

* When he was forced to deny that the earth went round the sun. Giordano Bruno had been burnt for this heresy in 1600.

perhaps lucky to be allowed to retire to his villa at Arcetri, near Florence. Living there, virtually under house-arrest, he took up the study of the strength of materials as being, I suppose, the safest and least subversive subject he could think of.

As it happened, Galileo's own contribution to our knowledge of the strength of materials was only moderately distinguished, though one must bear in mind that he was almost seventy when he began to work on the subject, that he had been through a great deal and that he was still more or less a prisoner. However, he was allowed to correspond with scholars in various parts of Europe, and his great reputation lent prestige and publicity to any subject he took up.

Among his many surviving letters there are several about structures, and his correspondence with Mersenne, who worked in France, seems to have been particularly fruitful. Marin Mersenne (1588–1648) was a Jesuit priest, but presumably nobody could object to his researches on the strength of metal wires. Edmé Mariotte (1620–84), a much younger man, was also a priest, being Prior of Saint Martin-sous-Beaune, near Dijon, in the wine country. He spent most of his life working on the laws of terrestrial mechanics and on the strength of rods in tension and in bending. Under Louis XIV he helped to found the French Academy of Sciences and was in favour with both Church and State. None of these people, it will be noted, were professional builders or shipwrights.

By Mariotte's time the whole subject of the behaviour of materials and structures under loads was beginning to be called the science of elasticity – for reasons which will become apparent in the next chapter – and we shall use this name repeatedly throughout this book. Since the subject became popular with mathematicians about 150 years ago I am afraid that a really formidable number of unreadable, incomprehensible books have been written about elasticity, and generations of students have endured agonies of boredom in lectures about materials and structures. In my opinion the mystique and mumbo-jumbo is overdone and often beside the point. It is true that the higher flights of elasticity are mathematical and very difficult – but then this sort of theory is probably only rarely used by successful engineering designers. What is actually needed for a great many ordinary purposes can be understood quite easily by any intelligent person who will give his or her mind to the matter.

The man in the street, or the man in the workshop, thinks he needs virtually no theoretical knowledge. The engineering don is apt to pretend that to get anywhere worth while without the higher mathematics is not only impossible but that it would be vaguely immoral if you could. It seems to me that ordinary mortals like you and me can get along surprisingly well with some intermediate – and I hope more interesting – state of knowledge.

All the same, we cannot wholly evade the question of mathematics, which is said to have originated in Babylonia – possibly at the time of the Tower of Babel incident. Mathematics is to the scientist and the engineer a tool, to the professional mathematician a religion, but to the ordinary person a stumbling-block. Yet all of us are really using mathematics through every moment of our lives. When we play tennis or walk downstairs we are actually solving whole pages of differential equations, quickly, easily and without thinking about it, using the analogue computer which we keep in our minds. What we find difficult about mathematics is the formal, symbolic presentation of the subject by pedagogues with a taste for dogma, sadism and incomprehensible squiggles.

For the most part, wherever a 'mathematical' argument is really needed I shall try to use graphs and diagrams of the simplest kind. We shall, however, need some arithmetic and a little very, very elementary algebra, which – however rude we may be to the mathematicians – is, after all, a simple, powerful and convenient mode of thought. Even if you are born, or think you are born, with an allergy to algebra, please do not be frightened of it. However, if you really must skip it, it will still be possible to follow the arguments in this book in a qualitative way without losing too much of the story.

One further point: structures are made from materials and we shall talk about structures and also about materials; but in fact there is no clear-cut dividing line between a material and a structure. Steel is undoubtedly a material and the Forth bridge is undoubtedly a structure, but reinforced concrete and wood and human flesh – all of which have a rather complicated constitution

– may be considered as either materials or structures. I am afraid that, like Humpty-Dumpty, when we use the word 'material' in this book, it will mean whatever we want it to mean. That this is not always the same as what other people mean by 'material' was brought home to me by another lady at another cocktail party.

'Do tell me what it is you do?'

'I'm a professor of materials.'

'What fun it must be to handle all those dress-fabrics!'

Part One

The difficult birth of the science of elasticity

Chapter 2   Why structures carry loads

or the springiness of solids

> *Let us begin at the beginning with Newton who said
> that action and reaction are equal and opposite. This
> means that every push must be matched and balanced
> by an equal and opposite push. It does not matter
> how the push arises. It may be a 'dead' load for
> instance: that is to say a stationary weight of some
> kind. If I weigh 200 pounds and stand on the floor,
> then the soles of my feet push downwards on the floor
> with a push or thrust of 200 pounds; that is the
> business of feet. At the same time the floor must push
> upwards on my feet with a thrust of 200 pounds; that
> is the business of floors. If the floor is rotten and
> cannot furnish a thrust of 200 pounds then I shall fall
> through the floor. If, however, by some miracle, the
> floor produced a larger thrust than my feet have
> called upon it to produce, say 201 pounds, then the
> result would be still more surprising because, of
> course, I should become airborne.*
>
> *The New Science of Strong Materials – or Why
> you don't fall through the floor* (Chapter 2)

We might start by asking how it is that any inanimate solid, such
as steel or stone or timber or plastic, is able to resist a mechanical
force at all – or even to sustain its own weight? This is, essentially,
the problem of 'Why we don't fall through the floor' and the
answer is by no means obvious. It lies at the root of the whole
study of structures and is intellectually difficult. In the event, it
proved too difficult for Galileo, and the credit for the achievement
of any real understanding of the problem is due to that very
cantankerous man Robert Hooke (1635–1702).

In the first place, Hooke realized that, if a material or a structure is to resist a load, it can only do so by pushing back at it with
an equal and opposite force. If your feet push down on the floor,
the floor must push up on your feet. If a cathedral pushes down

on its foundations, the foundations must push up on the cathedral. This is implicit in Newton's third law of motion, which, it will be remembered, is about action and reaction being equal and opposite.

In other words, a force cannot just get lost. Always and whatever happens every force must be balanced and reacted by another equal and opposite force at every point throughout a structure. This is true for any kind of structure, however small and simple or however large and complicated it may be. It is true, not only for floors and cathedrals, but also for bridges and aeroplanes and balloons and furniture and lions and tigers and cabbages and earthworms.

If this condition is not fulfilled, that is to say if all the forces are not in equilibrium or balance with each other, then either the structure will break or else the whole affair must take off, like a rocket, and end up somewhere in outer space. This latter result is frequently implicit in the examination answers of engineering students.

Let us consider for a moment the simplest possible sort of structure. Suppose that we hang a weight, such as an ordinary brick, from some support – which might be the branch of a tree – by means of a piece of string (Figure 1). The weight of the brick, like the weight of Newton's apple, is due to the effect of the earth's gravitational field upon its mass and it acts continually down-

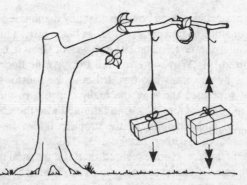

Figure 1. The weight of the brick, acting downwards, must be supported by an equal and opposite upward pull or tension in the string.

wards. If the brick is not to fall, then it must be sustained in its position in mid-air by a continuing equal and opposite upwards force or pull in the string. If the string is too weak, so that it cannot produce an upward force equal to the weight of the brick, then the string will break and the brick will fall to the ground – again like Newton's apple.

However, if our string is a strong one, so that we are able to hang not one, but two, bricks from it, then the string will now have to produce twice as much upward force; that is, enough to support both bricks. And so on, of course, for any other variations of the load. Moreover, the load does not have to be a 'dead' weight such as a brick; forces arising from any other cause, such as the pressure of the wind, must be resisted by the same sort of reaction.

In the case of the brick which hangs from a tree the load is supported by the tension in the string, in other words by a pull. In many structures, such as buildings, the load is carried in compression, that is by pushing. In both cases the general principles are the same. Thus if any structural system is to do its job – that is to say, if the load is supported in a satisfactory way so that nothing very much happens – then it must somehow manage to produce a push or a pull which is exactly equal and opposite to the force which is being applied to it. That is, it has to resist all the pushes and pulls which may happen to arrive upon its doorstep by pushing and pulling back at them by just the right amount.

This is all very well and it is generally fairly easy to see why a load pushes or pulls on a structure. The difficulty is to see why the structure should push or pull back at the load. As it happens, quite young children have had some inkling of the problem from time to time.

'Do stop pulling the cat's tail, darling.'

'I'm not pulling, Mummy, Pussy's pulling.'

In the case of the cat's tail the reaction is provided by the living biological activity of the cat's muscles pulling against the child's muscles, but of course this kind of active muscular reaction is not very often available, nor is it necessary.

If the cat's tail had happened to be attached, not to the cat, but to something inert, like a wall, then the wall would have to be doing the 'pulling'; whether the resistance to the child's pull is generated actively by the cat or passively by the wall makes no difference to the child or to the tail (Figures 2 and 3).

How then can an inert or passive thing like a wall or a string – or, come to that, a bone or a steel girder or a cathedral – produce the large reactive forces which are needed?

Figure 2. 'Do stop pulling the cat's tail, darling.'
'I'm not pulling, Mummy, Pussy's pulling.'
Figure 3. It doesn't make any difference whether Pussy pulls or not.

Hooke's law – or the springiness of solids

The power of any Spring is in the same proportion with the Tension* thereof: That is, if one power stretch or bend it one space, two will bend it two, three will bend it three, and so forward. And this is the Rule or Law of Nature, upon which all manner of Restituent or Springing motion doth proceed.
Robert Hooke

By about 1676 Hooke saw clearly that, not only must solids resist weights or other mechanical loads by pushing back at them, but also that

1. Every kind of solid *changes its shape* – by stretching or contracting itself – when a mechanical force is applied to it.

* In Hooke's time 'tension' meant what we should call 'extension', just as '*tensio*' did in Latin.

Figures 4 and 5. All materials and structures deflect, to greatly varying extents, when they are loaded. The science of elasticity is about the interactions between forces and deflections. The material of the bough is stretched near its upper surface and compressed or contracted near its lower surface by the weight of the monkey.

2. It is this change of shape which enables the solid to do the pushing back.

Thus, when we hang a brick from the end of a piece of string, the string gets longer, and it is just this stretching which enables the string to pull upwards on the brick and so prevent it from falling. *All* materials and structures deflect, although to greatly varying extents, when they are loaded (Figures 4 and 5).

It is important to realize that it is perfectly normal for any and every structure to deflect in response to a load. Unless this deflection is too large for the purposes of the structure, it is not in any way a 'fault' but rather an essential characteristic without which no structure would be able to work. *The science of elasticity is about the interactions between forces and deflections in materials and structures.*

Although *every* kind of solid changes its shape to some extent when a weight or other mechanical force is applied to it, the deflections which occur in practice vary enormously. With a thing like a plant or a piece of rubber the deflections are often very large and are easily seen, but when we put ordinary loads on hard substances like metal or concrete or bone the deflections are sometimes very small indeed. Although such movements are often far too small to see with the naked eye, they always exist and are perfectly real, even though we may need special appliances in order to measure them. When you climb the tower of a cathedral it becomes shorter, as a result of your added weight, by a very, very tiny amount, but it really does become shorter. As a matter of fact, masonry is really more flexible than you might think, as one can see by looking at the four principal columns which support the tower of Salisbury Cathedral: they are all quite noticeably bent (Plate 1).

Hooke made a further important step in his reasoning which, even nowadays, some people find difficult to follow. He realized that, when any structure deflects under load in the way we have been talking about, the material from which it is made is itself also stretched or contracted, internally, throughout all its parts and in due proportion, down to a very fine scale – as we know nowadays, down to a molecular scale. Thus, when we deform a stick or a steel spring – say by bending it – the atoms and molecules of which the material is made have to move further apart, or else squash closer together, when the material as a whole is stretched or compressed.

As we also know nowadays, the chemical bonds which join the atoms to each other, and so hold the solid together, are very strong and stiff indeed. So when the material as a whole is stretched or compressed this can only be done by stretching or compressing many millions of strong chemical bonds which vigorously resist being deformed, even to a very small extent. Thus these bonds produce the required large forces of reaction (Figure 6).

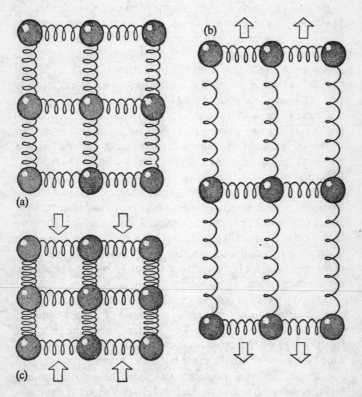

Figure 6.   Simplified model of distortion of interatomic bonds under mechanical strain.

   (a) Neutral, relaxed or strain-free position.

   (b) Material strained in tension, atoms further apart, material gets longer.

   (c) Material strained in compression, atoms closer together, material gets shorter.

Although Hooke knew nothing in detail about chemical bonds and not very much about atoms and molecules, he understood perfectly well that something of this kind was happening within the fine structure of the material, and he set out to determine what might be the nature of the macroscopic relationship between forces and deflections in solids.

He tested a variety of objects made from various materials and having various geometrical forms, such as springs and wires and beams. Having hung a succession of weights upon them and measured the resulting deflections, he showed that the deflection in any given structure was usually proportional to the load. That is to say, a load of 200 pounds would cause twice as much deflection as a load of 100 pounds 'and so forward'.

Furthermore, within the accuracy of Hooke's measurements – which was not very good – most of these solids recovered their original shape when the load which was causing the deflection was removed. In fact he could usually go on loading and unloading structures of this kind indefinitely without causing any permanent change of shape. Such behaviour is called 'elastic' and is common. The word is often associated with rubber bands and underclothes, but it is just as applicable to steel and stone and brick and to biological substances like wood and bone and tendon. It is in this wider sense that engineers generally use it. Incidentally, the 'ping' of the mosquito, for instance, is due to the highly elastic behaviour of the resilin springs which operate its wings.

However, a certain number of solids and near-solids, like putty and plasticine, do not recover completely but remain distorted when the load is taken off. This kind of behaviour is called 'plastic'. The word is by no means confined to the materials from which ashtrays are usually made but is also applied to clay and to soft metals. Such plastic substances shade off into things like butter and porridge and treacle. Furthermore, many of the materials which Hooke considered to be 'elastic' turn out to be imperfectly so when tested by more accurate modern methods.

However, as a broad generalization, Hooke's observations remain true and still provide the basis of the modern science of elasticity. Nowadays, and with hindsight, the idea that most materials and structures, not only machinery and bridges and buildings but also trees and animals and rocks and mountains and the round world itself, behave very much like springs may seem simple enough – perhaps blindingly obvious – but, from his diary, it is clear that to get thus far cost Hooke great mental effort and many doubts. It is perhaps one of the great intellectual achievements of history.

After he had tried out his ideas on Sir Christopher Wren in a series of private arguments, Hooke published his experiments in 1679 in a paper called 'De potentia restitutiva or of a spring'. This paper contained the famous statement '*ut tensio sic vis*' ('as the extension, so the force'). This principle has been known for three hundred years as 'Hooke's law'.

How elasticity got bogged down

But to make an enemy of Newton was fatal. For Newton, right or wrong, was implacable.

Margaret 'Espinasse, *Robert Hooke* (Heinemann, 1956)

Although in modern times Hooke's law has been of the very greatest service to engineers, in the form in which Hooke originally propounded it its practical usefulness was rather limited. Hooke was really talking about the deflections of a complete structure – a spring, a bridge or a tree – when a load is applied to it.

If we think for one moment, it is obvious that the deflection of a structure is affected *both* by its size and geometrical shape and *also* by the sort of material from which it is made. Materials vary very greatly in their intrinsic stiffness. Things like rubber or flesh are easily distorted by small forces which we can apply with our fingers. Other substances such as wood and bone and stone and most metals are very considerably stiffer, and, although no material can be absolutely 'rigid', a few solids like sapphire and diamond are very stiff indeed.

We can make objects of the same size and shape, such as ordinary plumber's washers, out of steel and also out of rubber. It is clear that the steel washer is very much more rigid (in fact about 30,000 times more rigid) than the rubber one. Again, if we make a thin spiral spring and also a thick and massive girder from the same material – such as steel – then the spring will naturally be very much more flexible than the girder. We need to be able to separate and to quantify these effects, for in engineering, as in biology, we are ringing the changes of these variables all the time and we need some reliable way of sorting the whole thing out.

After such a promising start it is rather surprising that no scientific way of coping with this difficulty emerged until 120 years after Hooke's death. In fact, throughout the eighteenth century remarkably little real progress was made in the study of elasticity. The reasons for this lack of progress were no doubt complex, but in general it can be said that, while the scientists of the seventeenth century saw their science as interwoven with the progress of technology – a vision of the purpose of science which was then almost new in history – many of the scientists of the eighteenth century thought of themselves as philosophers working on a plane which was altogether superior to the sordid problems of manufacturing and commerce. This was, of course, a reversion to the Greek view of science. Hooke's law provided a broad philosophical explanation of some rather commonplace phenomena which was quite adequate for the gentleman-philosopher who was not very interested in the technical details.

With all this, however, we cannot leave out the personal influence of Newton (1642–1727) himself or the after-effects of the bitter enmity which existed between Newton and Hooke. Intellectually, Hooke was probably nearly as able as Newton, and he was certainly even more touchy and vain; but in other respects they were men of totally different temperaments and interests. Basically, although they both came from fairly modest backgrounds, Newton was a snob whereas Hooke, though a personal friend of Charles II, was not.

Unlike Newton, Hooke was an earthy sort of person who was occupied with an enormous number of very practical problems about elasticity and springs and clocks and buildings and microscopes and the anatomy of the common flea. Among Hooke's inventions which are still in use today are the universal joint, used in car transmissions, and the iris diaphragm, which is used in most cameras. Hooke's carriage lamp, in which, as the candle burnt down, its flame was kept in the centre of the optical system by means of a spring feed, went out of use only in the 1920s. Such lamps are still to be seen outside people's front doors. Furthermore, Hooke's private life out-sinned that of his friend Samuel Pepys: not only was every servant girl fair game to him, but he lived for many years '*perfecte intime omne*'* with his attractive niece.

Newton's vision of the Universe may have been wider than Hooke's, but his interest in science was much less practical. In fact, like that of many lesser dons, it could often be described as anti-practical. It is true that Newton became Master of the Mint and did the job well, but it seems that his acceptance of the post had little to do with any desire to apply science and a lot to do with the fact that this was a 'place under Government' which, in those days, conferred a much higher social position than his fellowship of Trinity, not to mention a higher salary. A great deal of Newton's time, however, was spent in a curious world of his own in which he speculated about such perplexing theological problems as the Number of the Beast. I don't think he had much time or inclination to indulge in the sins of the flesh.

In short, Newton was well constituted to detest Hooke as a man and to loathe everything he stood for, down to and including elasticity. It so happened that Newton had the good fortune to live on for twenty-five years after Hooke died, and he devoted a good deal of this time to denigrating Hooke's memory and the importance of applied science. Since Newton had, by then, an almost God-like position in the scientific world, and since all this tended to reinforce the social and intellectual tendencies of the age, subjects like structures suffered heavily in popularity, even for many years after Newton's death.

Thus the situation throughout the eighteenth century was that, while the manner in which structures worked had been explained in a broad general way by Hooke, his work was not much followed up or exploited, and so the subject remained in such a condition that detailed practical calculations were scarcely possible.

So long as this state of affairs continued the usefulness of theoretical elasticity in engineering was limited. French eighteenth.

* Hooke's own phrase. Her name was Grace.

century engineers were aware of this but regretted it and tried to build structures (which quite often fell down) making use of such theory as was available to them. English engineers, who were also aware of it, were usually indifferent to 'theory' and they built the structures of the Industrial Revolution by rule-of-thumb 'practical' methods. These structures probably fell down nearly, but not quite, as often.

Chapter 3   The invention of stress and strain

*– or Baron Cauchy and the decipherment of
Young's modulus*

*What would life be without arithmetic, but a scene of
horrors?*

Rev. Sydney Smith, letter to a young lady, 22
July 1835

Apart from Newton and the prejudices of the eighteenth century,
the main reason why the science of elasticity got stuck for so long
was that the few scientists who did study it tried to deal with
forces and deflections by considering the structure as a whole – as
Hooke had done – rather than by analysing the forces and extensions which could be shown to exist at any given *point within* the
material. All through the eighteenth century and well into the
nineteenth, very clever men, such as Leonhard Euler (1707–83)
and Thomas Young (1773–1829), performed what must appear to
the modern engineer to be the most incredible intellectual contor-
tions in their attempts to solve what now seem to us to be quite
straightforward problems.

The concept of the elastic conditions at a specified point inside
a material is the concept of stress and strain. These ideas were
first put forward in a generalized form by Augustin Cauchy (1789–
1857) in a paper to the French Academy of Sciences in 1822. This
paper was perhaps the most important event in the history of
elasticity since Hooke. After this, that science showed promise of
becoming a practical tool for engineers rather than a happy
hunting-ground for a few somewhat eccentric philosophers. From
his portrait, painted at about this time, Cauchy looks rather a pert
young man, but he was undoubtedly an applied mathematician of
great ability.

When, eventually, English nineteenth-century engineers
bothered to read what Cauchy had said on the subject, they found
that, not only were the basic concepts of stress and strain really
quite easy to understand, but, once they had been understood, the

whole study of structures was much simplified. Nowadays these ideas can be understood by anybody,* and it is hard to account for the bewildered and even resentful attitude which is sometimes taken up by laymen when 'stresses and strains' are mentioned. I once had a research student with a nice new degree in zoology who was so upset by the whole idea of stress and strain that she ran away from the university and hid herself. I still do not see why.

Stress – which is not to be confused with strain

As it happened, Galileo himself very nearly stumbled upon the idea of stress. In the *Two New Sciences*, the book he wrote in his old age at Arcetri, he states very clearly that, other things being equal, a rod which is pulled in tension has a strength which is proportional to its cross-sectional area. Thus, if a rod of two square centimetres cross-section breaks at a pull of 1,000 kilograms, then one of four square centimetres cross-section will need a pull of 2,000 kilograms force in order to break it, and so on. That it should have taken nearly two hundred years to divide the breaking load by the area of the fracture surface, so as to get what we should now call a 'breaking stress' (in this case 500 kilograms per square centimetre) which might be applied to all similar rods made from the same material almost passes belief.

Cauchy perceived that this idea of stress can be used, not only to predict when a material will break, but also to describe the state of affairs at any point inside a solid in a much more general kind of way. In other words the 'stress' in a solid is rather like the 'pressure' in a liquid or a gas. It is a measure of how hard the atoms and molecules which make up the material are being pushed together or pulled apart as a result of external forces.

Thus, to say 'The stress at that point in this piece of steel is 500 kilograms per square centimetre' is no more obscure or mysterious than to say 'The pressure of the air in the tyres of my car is 2 kilograms per square centimetre – or 28 pounds per square

---

*Except, apparently, the *Oxford Dictionary*. The words are used, of course, in casual conversation to describe the mental state of people and as if they meant the same thing. In physical science the meanings of the two words are quite clear and distinct.

inch.' However, although the concepts of pressure and stress are fairly closely comparable, we have to bear in mind that pressure acts in all three directions within a fluid while the stress in a solid is often a directional or one-dimensional affair. Or, at any rate, so we shall consider it for the present.

Numerically, the stress in any direction at a given point in a material is simply the force or load which happens to be acting in that direction at the point, divided by the area on which the force acts.* If we call the stress at a certain point s, then

$$\text{Stress} = s = \frac{\text{load}}{\text{area}} = \frac{P}{A}$$

where P = load or force and A is the area over which the force P can be considered as acting.

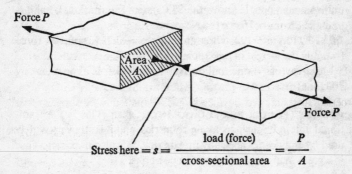

Force P

Area
A

Force P

$$\text{Stress here} = s = \frac{\text{load (force)}}{\text{cross-sectional area}} = \frac{P}{A}$$

Figure 1.   Stress in a bar under tension. (Compressive stress is exactly analogous.)

To revert to our brick, which we left in the last chapter hanging from its string. If the brick weighs 5 kilograms and the string has a cross-section of 2 square millimetres, then the brick pulls on the string with a force of 5 kilograms, and the stress in the string will be:

*How can a 'point' have an 'area'? Consider the analogy of speed: we express speed as the distance covered in a certain length of time, e.g. miles per hour, although we are concerned usually with the speed at any given – infinitely brief – moment.

$$s = \frac{\text{load}}{\text{area}} = \frac{P}{A} = \frac{5 \text{ kilograms force}}{2 \text{ square millimetres}}$$
$$= 2 \cdot 5 \text{ kilograms force per square millimetre}$$

or, if we prefer it, 250 kilograms force per square centimetre or kgf/cm².

Units of stress

This raises the vexed question of units of stress. Stress can be expressed in any units of force divided by any units of area – and it frequently is. To reduce the amount of confusion we shall stick to the following units in this book.

MEGANEWTONS PER SQUARE METRE: MN/m². This is the SI unit. As most people know, the SI (System International) habit is to make the unit of force the Newton.
1·0 Newton = 0·102 kilograms force = 0·225 pounds force (roughly the weight of one apple).
1 Meganewton = one million Newtons, which is almost exactly 100 tons force.

POUNDS (FORCE) PER SQUARE INCH: p.s.i. This is the traditional unit in English-speaking countries, and it is still very widely used by engineers, especially in America. It is also in common use in a great many tables and reference books.

KILOGRAMS (FORCE) PER SQUARE CENTIMETRE: kgf/cm² (sometimes kg/cm²). This is the unit in common use in Continental countries, including Communist ones.

FOR CONVERSION

$$1 \text{ MN/m}^2 = 10 \cdot 2 \text{ kgf/cm}^2 = 146 \text{ p.s.i.}$$
$$1 \text{ p.s.i.} = 0 \cdot 00685 \text{ MN/m}^2 = 0 \cdot 07 \text{ kgf/cm}^2$$
$$1 \text{ kgf/cm}^2 = 0 \cdot 098 \text{ MN/m}^2 = 14 \cdot 2 \text{ p.s.i.}$$

Thus the stress in our piece of string, which we found to be 250 kgf/cm², is also equal to 24·5 MN/m² or 3,600 p.s.i. Since the calculation of stresses is not usually a very accurate business, there is no sense in fussing too much about very exact conversion factors.

It is worth repeating that it is important to realize that the stress in a material, like the pressure in a fluid, is a condition which exists at a *point* and it is not especially associated with any particular cross-sectional area, such as a square inch or a square centimetre or a square metre.

Strain – which is not the same thing as stress

Just as stress tells us how *hard* – that is, with how much force – the atoms at any point in a solid are being pulled apart, so strain tells us how *far* they are being pulled apart – that is, by what proportion the bonds between the atoms are stretched.

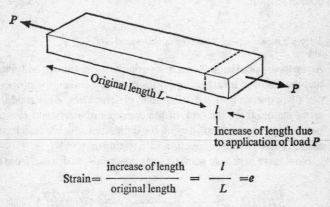

$$\text{Strain} = \frac{\text{increase of length}}{\text{original length}} = \frac{l}{L} = e$$

Figure 2.   Strain in a bar under tension. (Compressive strain is exactly analogous.)

Thus, if a rod which has an original length L is caused to stretch by an amount l by the action of a force on it, then the *strain*, or proportionate change of length, in the rod will be e, let us say, such that:

$$e = \frac{l}{L}$$

To return to our string, if the original length of the string was, say, 2 metres (or 200 cm), and the weight of the brick causes it to stretch by 1 centimetre, then the strain in the string is:

$$e = \frac{l}{L} = \frac{1}{200} = 0.005 \text{ or } 0.5\%$$

Engineering strains are usually quite small, and so engineers very often express strains as percentages, which reduces the opportunities for confusion with noughts and decimal points.

Like stress, strain is not associated with any particular length or cross-section or shape of material. It is also a condition at a point. *Again, since we calculate strain by dividing one length by another length – i.e. the extension by the original length – strain is a ratio, which is to say a number, and it has no units, S I, British or anything else.* All this applies just as much in compression, of course, as it does in tension.

Young's modulus – or how stiff is this material?

As we have said, Hooke's law in its original form, though edifying, was the result of a rather inglorious muddle between the properties of materials and the behaviour of structures. This muddle arose mainly from the lack of the concepts of stress and strain, but we also have to bear in mind the difficulties which would have existed in the past in connection with testing materials.

Nowadays, when we want to test a material – as distinct from a

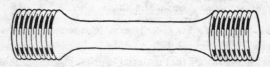

Figure 3. A typical tensile test-piece.

structure – we generally make what is called a 'test-piece' from it. The shapes of test-pieces may vary a good deal but usually they have a parallel stem, on which measurements can be made, and are provided with thickened ends by which they can be attached to the testing machine. An ordinary metal test-piece often looks like Figure 3.

Testing machines also vary a good deal in size and in design, but basically they are all mechanical devices for applying a measured load in tension or in compression.

The stress in the stem of the test-piece is obtained merely by dividing the load recorded at each stage on the dial of the machine by the area of its cross-section. The extension of the stem of the test-piece under load – and therefore the strain in the material – is usually measured by means of a sensitive device called an extenso-meter, which is clamped to two points on the stem.

With equipment of this kind it is generally quite easy to measure the stress and the strain which occur within a specimen of a

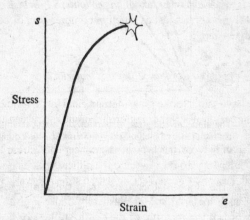

Figure 4.   A typical 'stress-strain diagram'.

material as we increase the load upon it. The relationship between stress and strain for that material is given by the graph of stress plotted against strain which we call the 'stress-strain diagram'. This stress-strain diagram, which may look something like Figure 4, is very characteristic of any given material, and its shape is usually unaffected by the size of the test-piece which happens to have been used.

When we come to plot the stress-strain diagram for metals and for a number of other common solids we are very apt to find that, at least for moderate stresses, the graph is a straight line. When this is so we speak of the *material* as 'obeying Hooke's law' or sometimes of a 'Hookean material'.

What we also find, however, is that the *slope* of the straight part of the graph varies greatly for different materials (Figure 5). It is

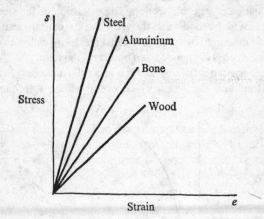

Figure 5. The slope of the straight part of the stress-strain diagram is characteristic of each different material. *E*, the Young's modulus of elasticity, represents this slope.

clear that the slope of the stress-strain diagram measures how readily each material strains elastically under a given stress. In other words it is a measure of the elastic stiffness or floppiness of a given solid.

For any given material which obeys Hooke's law, the slope of the graph or the ratio of stress to strain will be constant. Thus for any particular material

$$\frac{\text{stress}}{\text{strain}} = \frac{s}{e} = \text{Young's modulus of elasticity, which we call } E$$

$$= \text{constant for that material}$$

Young's modulus is sometimes called 'the elastic modulus' and sometimes '*E*', and is quite often spoken of as 'stiffness' in ordinary technical conversation. The word 'modulus', by the way, is Latin for 'a little measure'.

Our string, it may be remembered, was strained 0·5 per cent or 0·005 by the weight of the brick, which imposed a stress of 24·5 MN/m² or 3,600 p.s.i. The Young's modulus of the string is therefore

$$\frac{\text{stress}}{\text{strain}} = \frac{24\cdot5}{0\cdot005} = 4{,}900 \text{ MN/m}^2$$

$$= 720{,}000 \text{ p.s.i.}$$

Units of stiffness or Young's modulus

Since we are dividing a stress by a fraction, which is to say a number, which has no dimensions, Young's modulus has the same dimensions as a stress and is expressed in stress units, that is to say MN/m², p.s.i. or kgf/cm². Since, however, Young's modulus may be regarded as that stress which would double the length of the material (i.e. the stress at 100 per cent strain) – if the material did not break first – the numbers involved are often large, and some people find them difficult to visualize.

Practical values of Young's modulus

The Young's moduli of a number of common biological and engineering materials are given in Table 1. Starting from the cuticle of the pregnant locust (which is low, but not very exceptionally low, for biological materials; the cuticle of the male locust and of the virgin female locust is a lot stiffer, by the way) the Young's moduli are arranged in ascending order all the way to diamond. It will be seen that the range of stiffness varies by about 6,000,000 to one. Which is a lot. We shall discuss why this should be so in Chapter 8.

It may be noticed that a good many common soft biological materials do not occur in this table. This is because their elastic behaviour does not obey Hooke's law, even approximately, so that it is really impossible to define a Young's modulus, at any

TABLE 1

Approximate Young's moduli of various solids

Material	Young's modulus (E)	
	p.s.i.	MN/m²
Soft cuticle of pregnant locust*	30	0.2
Rubber	1,000	7
Shell membrane of egg	1,100	8
Human cartilage	3,500	24
Human tendon	80,000	600
Wallboard	200,000	1,400
Unreinforced plastics, polythene, nylon	200,000	1,400
Plywood	1,000,000	7,000
Wood (along grain)	2,000,000	14,000
Fresh bone	3,000,000	21,000
Magnesium metal	6,000,000	42,000
Ordinary glasses	10,000,000	70,000
Aluminium alloys	10,000,000	70,000
Brasses and bronzes	17,000,000	120,000
Iron and steel	30,000,000	210,000
Aluminium oxide (sapphire)	60,000,000	420,000
Diamond	170,000,000	1,200,000

* By courtesy of Dr Julian Vincent, Department of Zoology, University of Reading.

rate in the terms we have been talking about. We shall come back to this sort of elasticity later on.

Young's modulus is nowadays regarded as a pretty fundamental concept; it thoroughly pervades engineering and materials science and is beginning to invade biology. Yet it took all of the first half of the nineteenth century for the penny to drop in the minds of engineers. This was partly due to sheer conservatism and partly due to the late arrival of any workable concept of stress and strain.

Given these ideas, few things are simpler or more obvious than Young's modulus; without them, the whole affair must have seemed impossibly difficult. Young, who was to play an important part in the decipherment of Egyptian hieroglyphics and who had one of the finest brains of his generation, obviously had a very severe intellectual struggle.

Working around the year 1800, he had to approach the problem by a route quite different from that which we have just used, and he considered the question in terms of what we should now call the 'specific modulus', that is, by how much a column of a material might be expected to shorten under its own weight. Young's own definition of his modulus, published in 1807, is as follows: 'The modulus of the elasticity of any substance is a column of the same substance, capable of producing a pressure on its base which is to the weight causing a certain degree of compression as the length of the substance is to the diminution of its length.'*

After which, Egyptian hieroglyphics must have appeared simple.

It was said of Young by one of his contemporaries that 'His words were not those in familiar use, and the arrangement of his ideas seldom the same as those he conversed with. He was therefore worse calculated than any man I ever knew for the communication of knowledge.' All the same we have to realize that Young was wrestling with an idea that was scarcely capable of expression without the concepts of stress and strain, which did not come into use until fifteen or twenty years later. The modern definition of Young's modulus ($E = $ stress/strain) was given in 1826 – three years before Young died – by the French engineer Navier (1785–1836). As the inventor of stress and strain, Cauchy was eventually made a baron by the French government. He seems to have deserved it.

Strength

It is necessary to avoid confusion between the strength of a structure and the strength of a material. The strength of a *structure* is simply the *load* (in pounds force or Newtons or kilograms force) which will just break the structure. This figure is known as the 'breaking load', and it naturally applies only to some individual, specific structure.

*'Though science is much respected by their Lordships and your paper is much esteemed, it is too learned . . . in short it is not understood' (Admiralty letter to Young).

The strength of a *material* is the *stress* (in p.s.i. or MN/m² or kgf/cm²) required to break a piece of the material itself. It will generally be the same for all specimens of any given solid. We are most often concerned with the tensile strength of materials, which is sometimes called the 'ultimate tensile stress' or U.T.S. This is usually determined by breaking small test-pieces in a testing machine. Naturally, the object of many strength calculations is to

TABLE 2

Approximate tensile strengths of various solids

Material	Tensile strength	
	p.s.i.	MN/m²
Non-metals		
Muscle tissue (fresh but dead)	15	0·1
Bladder wall (,, ,, ,,)	34	0·2
Stomach wall (,, ,, ,,)	62	0·4
Intestine (,, ,, ,,)	70	0·5
Artery wall (,, ,, ,,)	240	1·7
Cartilage (,, ,, ,,)	430	3·0
Cement and concrete	600	4·1
Ordinary brick	800	5·5
Fresh skin	1,500	10·3
Tanned leather	6,000	41·1
Fresh tendon	12,000	82
Hemp rope	12,000	82
Wood (air dry): along grain	15,000	103
across grain	500	3·5
Fresh bone	16,000	110
Ordinary glass	5,000–25,000	35–175
Human hair	28,000	192
Spider's web	35,000	240
Good ceramics	5,000–50,000	35–350
Silk	50,000	350
Cotton fibre	50,000	350
Catgut	50,000	350
Flax	100,000	700
Fibreglass plastics	50,000–150,000	350–1,050
Carbon-fibre plastics	50,000–150,000	350–1,050
Nylon thread	150,000	1,050

Material	Tensile strength	
	p.s.i.	MN/m²
Metals		
STEELS		
Steel piano wire (very brittle)	450,000	3,100
High tensile engineering steel	225,000	1,550
Commercial mild steel	60,000	400
WROUGHT IRON		
Traditional	15,000–40,000	100–300
CAST IRON		
Traditional (very brittle)	10,000–20,000	70–140
Modern	20,000–40,000	140–300
OTHER METALS		
Aluminium: cast	10,000	70
wrought alloys	20,000–80,000	140–600
Copper	20,000	140
Brasses	18,000–60,000	120–400
Bronzes	15,000–80,000	100–600
Magnesium alloys	30,000–40,000	200–300
Titanium alloys	100,000–200,000	700–1,400

predict the strength of a structure from the known strength of its material.

The tensile strengths of a good many materials are given in Table 2. As with stiffness, it will be seen that the range of strengths in both biological and engineering solids is very wide indeed. For instance, the contrast between the weakness of muscle and the strength of tendon is striking, and this accounts for the very different cross-sections of muscles and their equivalent tendons. Thus the thick and sometimes bulging muscle in our calves transmits its tension to the bone of our heel, so that we can walk and jump, by means of the Achilles or calcaneal tendon, which, although it is pencil-thin, is generally quite adequate for the job. Again, we can see why engineers are unwise to put tensile forces on concrete unless that weak material is sufficiently reinforced with strong steel bars.

The strong metals are rather stronger, on the whole, than the strong non-metals. However, nearly all metals are considerably denser than most biological materials (steel has a specific gravity of 7·8, most zoological tissues about 1·1). Thus, strength for weight, metals are not too impressive when compared with plants and animals.

We might now sum up what has been said in this chapter:

$$Stress = \frac{load}{area}$$

It expresses how *hard* (i.e. with how much force) the atoms at a point within a solid are being pulled apart or pushed together by a load.

$$Strain = \frac{extension\ under\ load}{original\ length}$$

It expresses how *far* the atoms at a point within a solid are being dragged apart or pushed together.
Stress is not the same thing as strain.
Strength. By the *strength* of a material we usually mean that stress which is needed to break it.

$$Young's\ modulus = \frac{stress}{strain} = E.$$

It expresses how stiff or how floppy a material is.
Strength is not the same thing as stiffness.

To quote from *The New Science of Strong Materials*: 'A biscuit is stiff but weak, steel is stiff and strong, nylon is flexible (low E) and strong, raspberry jelly is flexible (low E) and weak. The two properties together describe a solid about as well as you can reasonably expect two figures to do.'

In case you should ever have felt any trace of doubt or confusion on these points, it might be of some comfort to know that, not so long ago, I spent a whole evening in Cambridge trying to explain to two scientists of really shattering eminence and world-wide fame the basic difference between stress and strain and strength and stiffness in connection with a very expensive project about which they were proposing to advise the government. I am still uncertain how far I was successful.

Chapter 4   Designing for safety

– or can you really trust strength calculations?

That with music loud and long,
I would build that dome in air,
That sunny dome! Those caves of ice!
And all who heard should see them there,
And all should cry, Beware! Beware!

S. T. Coleridge, *Kubla Khan*

Naturally all this business about stresses and strains is only a means to an end; that is, to enable us to design safer and more effective structures and devices of one kind or another and to understand better how such things work.

Apparently Nature does not have to bother. The lilies of the field toil not, neither do they calculate, but they are probably excellent structures, and indeed Nature is generally a better engineer than man. For one thing she has more patience and, for another, her way of going about the design process is quite different.

In living creatures the broad general arrangement or lay-out of the parts is controlled during growth by the R N A–D N A mechanism – the famous 'double helix' of Wilkins, Crick and Watson.* However, in each individual plant or animal, once the general arrangement has been achieved, there is a good deal of latitude about the structural details. Not only the thickness, but also the composition of each load-carrying component is determined, to a considerable extent, by the use which is actually made of it and by the forces which it has to resist during life.† Thus the proportions of a living structure tend to become optimized with regard to its strength. Nature seems to be a pragmatic rather than a

---

*See, for instance, *The Double Helix*, by James D. Watson, Weidenfeld & Nicolson, 1968.

†The process also works in reverse; the bones of astronauts lose calcium and become weaker after a period of weightlessness in space.

mathematical designer; and, after all, bad designs can always be eaten by good ones.

Unfortunately, these design methods are not, as yet, available to human engineers, who are therefore driven to use either guess-work or calculation or, more often, some combination of the two. Both for safety and for economy it is clearly desirable to be able to predict how the various parts of an engineering structure will share the load between them and so to determine how thick or how thin they ought to be. Again, we generally want to know what deflections to expect when a structure is loaded, because it may be just as bad a thing for a structure to be too flexible as for it to be too weak.

French theory versus British pragmatism

Once the basic concepts of strength and stiffness had been stated and understood, a considerable number of mathematicians set themselves to devise techniques for analysing elastic systems oper-ating in two and three dimensions, and they began to use these methods to examine the behaviour of many different shapes of structures under loads. It happened that, during the first half of the nineteenth century, most of these theoretical elasticians were Frenchmen. Although very possibly there is something about elasticity which is peculiarly suited to the French temperament,* the practical encouragement for this research seems to have come, directly or indirectly, from Napoleon I and from the École Poly-technique, which was founded in 1794.

Because much of this work was abstract and mathematical it was not understood or generally accepted by most practising engineers until about 1850. This was especially the case in England and America, where practical men were regarded as greatly superior to 'mere theoreticians'. And besides, one Englishman had always beaten three Frenchmen. Of the Scottish engineer,

---

* Almost the only woman to have gained distinction in elasticity, Mademoi-selle Sophie Germain (1776–1831), was French. It may be relevant that two of our most highly educated and theoretically-minded engineers during this period, Sir Marc Brunel (1769–1849) and his son, Isambard Kingdom Brunel (1806–59), were of French origin.

Thomas Telford (1757–1834), whose magnificent bridges we can still admire, it is related that:

He had a singular distaste for mathematical studies, and never even made himself acquainted with the elements of geometry; so remarkable indeed was this peculiarity that when we had occasion to recommend to him a young friend as a neophyte in his office, and founded our recommendation on his having distinguished himself in mathematics, he did not hesitate to say that he considered such acquirements as rather disqualifying than fitting him for the situation.

Telford, however, really was a great man, and, like Nelson, he tempered his confidence with an attractive humility. When the heavy chains for the Menai suspension bridge (Plate 11) had been hoisted successfully in the presence of a large crowd, Telford was discovered, away from the cheering spectators, giving thanks on his knees.†

Not all engineers were as inwardly humble as Telford, and Anglo-Saxon attitudes at this time were often tinged, not only with intellectual idleness, but also with arrogance. Even so, scepticism about the trustworthiness of strength calculations was not unjustified. We must be clear that what Telford and his colleagues were objecting to was not a numerate approach as such – they were at least as anxious as anybody else to know what forces were acting on their materials – but rather the means of arriving at these figures. They felt that theoreticians were too often blinded by the elegance of their methods to the neglect of their assumptions, so that they produced the right answer to the wrong sum. In other words, they feared that the arrogance of mathematicians might be more dangerous than the arrogance of pragmatists, who, after all, were more likely to have been chastened by practical experience.

Shrewd North-Country consulting engineers realized, as all successful engineers must, that when we analyse a situation mathe-

† The British tradition of totally ignoring mathematics has been splendidly continued in the present century by a number of distinguished engineers, notably Sir Henry Royce, who did, after all, create the 'best car in the world'.

matically, we are really making for ourselves an artificial working model of the thing we want to examine. We hope that this algebraical analogue or model will perform in a way which resembles the real thing sufficiently closely to widen our understanding and to enable us to make useful predictions.

With fashionable subjects like physics or astronomy the correspondence between model and reality is so exact that some people tend to regard Nature as a sort of Divine Mathematician. However attractive this doctrine may be to earthly mathematicians, there are some phenomena where it is wise to use mathematical analogies with great caution. The way of an eagle in the air; the way of a serpent upon a rock; the way of a ship in the midst of the sea and the way of a man with a maid are difficult to predict analytically. One does sometimes wonder how mathematicians ever manage to get married. After King Solomon had built his temple, he would probably have added that the way of a structure with a load has a good deal in common at least with ships and eagles.

The trouble with things like these is that many of the real situations which are apt to arise are so complicated that they cannot be fully represented by one mathematical model. With structures there are often several alternative possible modes of failure. Naturally the structure breaks in whichever of these ways turns out to be the weakest – which is too often the one which nobody had happened to think of, let alone do sums about.

A deep, intuitive appreciation of the inherent cussedness of materials and structures is one of the most valuable accomplishments an engineer can have. No purely intellectual quality is really a substitute for this. Bridges designed upon the best 'modern' theories by Polytechniciens like Navier sometimes fell down. As far as I know, none of the hundreds of bridges and other engineering works which Telford built in the course of his long professional life ever gave serious trouble. Thus, during the period when French structural theory was outstanding, a great proportion of the railways and bridges on the Continent were being built by gritty and taciturn English and Scottish engineers who had little respect for the calculus.

Factors of safety and factors of ignorance

All the same, after about 1850 even British and American engineers did begin to do calculations about the strength of important structures, such as large bridges. They calculated the highest probable tensile stresses in the structure by the methods of the day, and they saw to it that these stresses were less than the official 'tensile strength' of the material. To make quite sure, they made the highest calculated working stress *much* less – three or four or even seven or eight times less – than the strength of the material as determined by breaking a simple, smooth, parallel-stemmed testpiece.* This was called 'applying a factor of safety'. Any attempt to save weight and cost by reducing the factor of safety was only too likely to lead to disaster.

Accidents were very apt to be put down as due to 'defective material', and a few of them may have been. Metals, of course, do vary in strength between different samples, and there is always some risk of poor material being built into a structure. However, iron and steel usually vary in strength by only a few per cent and very, very rarely by anything like a factor of three or four, let alone seven or eight. Practically always discrepancies as big as this between the theoretical and the actual strengths are due to other causes; at some unknown place in the structure the real stress must be very much higher than the calculated stress, and thus the 'factor of safety' is sometimes referred to as the 'factor of ignorance'.

Nineteenth-century engineers usually made things which were subject to tension stresses, such as boilers and beams and ships, out of wrought iron or mild steel, which had, with some justice, the reputation of being 'safe' materials. When a large factor of ignorance had been applied to the strength calculations, such structures often turned out to be quite satisfactory, although in fact accidents continued to occur fairly frequently.

Trouble became increasingly common with ships. The demand for speed and lightness led both the Admiralty and the shipbuilders into difficulties, since ships tended to break in two at sea

---

*Factors of safety as high as eighteen were used in the design of connecting-rods for steam locomotives at least as late as 1910.

although the highest calculated stresses seemed to be quite safe and moderate. In 1901, for instance, a brand-new turbine destroyer, H.M.S. *Cobra*, one of the fastest ships in the world, suddenly broke in two and sank in the North Sea in fairly ordinary weather. Thirty-six lives were lost. Neither the subsequent court martial nor the Admiralty Committee of Inquiry shed much light on the technical causes of the accident.

In 1903, therefore, the Admiralty made and published a number of experiments with a similar destroyer, H.M.S. *Wolf*, at sea in rough weather. These showed that the stresses deduced from strain measurements made on the hull under real conditions were rather less than those calculated by the designers before the ship was built. Since both sets of stresses were far below the known 'strength' of the steel from which the ship was constructed – the factor of safety being between five and six – these experiments were only moderately enlightening.

Stress concentrations – or how to start a crack

The first important step towards the understanding of problems of this kind was achieved, not by very expensive practical experiments on full-scale structures, but by theoretical analysis. In 1913 C. E. Inglis, who was later Professor of Engineering at Cambridge and was the very opposite of a 'remote and ineffectual don', published a paper in the *Transactions of the Institution of Naval Architects* whose consequences and applications extend far beyond the strength of ships.

What Inglis said about elasticians was really very much what Lord Salisbury is supposed to have said about politicians, namely that it is a great mistake to use only small-scale maps. For nearly a century elasticians had been content to plot the distribution of stresses in broad, general or Napoleonic terms. Inglis showed that this approach can be relied on only when the material and the structure have smooth surfaces and no sudden changes of shape.

Geometrical irregularities, such as holes and cracks and sharp corners, which had previously been ignored, may raise the local stress – often only over a very small area – very dramatically indeed. Thus holes and notches may cause the stress in their immediate vicinity to be much higher than the breaking stress of the material, even when the general level of stress in the surrounding neighbourhood is low and, from general calculations, the structure might appear to be perfectly safe.

This fact had been known, of course, in a general kind of way, to the people who put the grooves in slabs of chocolate and to those who perforate postage stamps and other kinds of paper. A dressmaker cuts a 'nick' in the selvedge of a piece of cloth before she tears it. Serious engineers, however, had not shown much

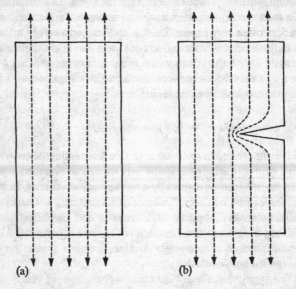

Figure 1. Stress trajectories in a bar uniformly loaded in tension (a) without and (b) with a crack.

interest in these fracture phenomena, which were not considered to belong to 'proper' engineering.

That almost any hole or crack or re-entrant in an otherwise continuous solid will cause a local increase of stress is easily explained. Figure 1a shows a smooth, uniform bar or plate of material, subject to a uniform tensile stress, *s*. The lines crossing the material represent what are called 'stress trajectories', that is to say, typical paths by which the stress is handed on from one molecule to the next. In this case they are, of course, straight parallel lines, uniformly spaced.

If we now interrupt a number of these stress trajectories by making a cut or a crack or a hole in the material, then the forces which the trajectories represent have to be balanced and reacted in some way. What actually happens is more or less what one would expect; the forces have to go round the gap, and as they do so the stress trajectories are crowded together to a degree which depends chiefly upon the shape of the hole (Figure 1b). In the case of a long crack, for instance, the crowding around the tip of the crack is often very severe. Thus in this immediate region there is more force per unit area and so the local stress is high (Plate 2).

Inglis was able to calculate the increase of stress which occurs at the tip of an elliptical hole in a solid which obeys Hooke's law.* Although his calculations are strictly true only for elliptical holes they apply with sufficient accuracy to openings of other shapes. Thus they apply not only to port-holes and doors and hatchways in ships and aeroplanes and similar structures but also to cracks and scratches and holes in all sorts of other materials and devices – to fillings in teeth, for example.

In terms of simple algebra what Inglis said was that, if we have a piece of material which is subject to a remotely applied stress s, and if we make a notch or a crack or a re-entrant of any kind in it having a length or depth L, and if this crack or re-entrant has a radius at the tip of r, then the stress at and very near to the tip is no longer s but is raised to:

$$s\left(1+2\sqrt{\frac{L}{r}}\right)$$

For a semi-circular notch or a round hole (when $r = L$) the stress will thus have the value of $3s$; but for openings like doors and hatchways, which often have sharp corners, r will be small

---

* As a matter of fact the effect of a round hole in a plate under tension had been calculated by Kirsch in Germany in 1898 and that of an elliptical hole by Kolosoff in Russia in 1910, but, as far as I know, little notice was taken of these results in English shipbuilding circles.

and L large, and so the stress at the corners may be very high – quite high enough to account for ships breaking in two.

In the *Wolf* experiments, extensometers, or strain-gauges, were clamped to the ship's plating in various positions. By this means the extension or elastic movement of the steel plates could be read off. From this the strain – and thus the stress – in the steel was easily calculated. As it happened none of the extensometers was placed close to the corners of hatchways or other openings. If this had been done some very frightening readings would almost certainly have been obtained when the ship was plunging into a head sea in Portland Race.

When we turn from hatchways to cracks the situation is even worse, because, while cracks are often centimetres or even metres long, the radius of the tip of the crack may be of molecular dimensions – less than a millionth of a centimetre – so that $\sqrt{L/r}$ is very large; thus the stress at the tip of the crack may well be a hundred or even a thousand times higher than the stress elsewhere in the material.

If Inglis's results had to be taken entirely at their face value it would scarcely be possible to make a safe tension structure at all. In fact the materials which are actually used in tension, metals, wood, rope, Fibreglass, textiles and most biological materials, are 'tough', which means, as we shall see in the next chapter, that they contain more or less elaborate defences against the effects of stress concentrations. However, even in the best and toughest of materials, this protection is only relative, and every tension structure is susceptible to some extent.

The 'brittle solids', however, which are used in technology, like glass and stone and concrete, do not have these defences. In other words they correspond pretty closely to the assumptions which were made in Inglis's calculations. Moreover we do not need to put in stress-raising notches artificially in order to weaken these materials. Nature has already done this liberally, and real solids are nearly always full of all kinds of small holes and cracks and scratches, even before we begin to make a structure out of them.

For these reasons it is rash to use any of the brittle solids in situations where they may be subject to appreciable tension stresses. They are, of course, very widely used in masonry and for roads and so on where they are, at least officially, in compression. Where we cannot avoid a certain amount of tension, for instance in glass windows, we have to take care to keep the tensile stresses very small indeed and to use a large factor of safety.

In talking of stress concentrations we must note that weakening effects are not exclusively caused by holes and cracks and other deficiencies of material. One can also cause stress concentrations by adding material, if this induces a sudden local increase of stiffness. Thus if we put a new patch on an old garment or a thick plate of armour on the thin side of a warship, no good will come of it.*

The reason for this is that the stress trajectories are diverted just as much by an area which strains too little, such as a stiff patch, as they are by an area which strains too much, such as a hole. Anything which is, so to speak, elastically out of step with the rest of the structure will cause a stress concentration and may therefore be dangerous.

When we seek to 'strengthen' something by adding extra material we have to be careful we do not in fact make it weaker. The inspectors employed by insurance companies and government departments who insist on pressure vessels and other structures being 'strengthened' by the addition of extra gussets and webs are sometimes responsible, in my experience, for the very accidents which they have tried to prevent.

Nature is generally rather good at avoiding stress concentrations of this and other kinds. However, one would think that stress concentrations must be of significance in orthopaedic surgery, especially when the surgeon fits a stiff metal prosthesis to a relatively flexible bone.

NOTE. In Inglis's formula (p. 67) L is the length of a crack proceeding inwards from the surface, i.e. half the length of an internal crack.

* 'Partial strength produces general weakness' (Sir Robert Seppings (1767–1840), Surveyor of the Navy 1813–32).

Chapter 5   Strain energy and modern fracture mechanics

– with a digression on bows, catapults and kangaroos

An unwise man doth not well consider this: and a fool doth not understand it.

Psalm 92

As we said in the last chapter, it was the considerable achievement of the nineteenth-century mathematicians to find ways of calculating the distribution and the magnitude of the stresses in most kinds of structures in a rather broad, generalized or academic way. However, many practical engineers had not long come to terms with calculations of this kind before Inglis planted the seeds of doubt at the back of their minds. Using the elasticians' own algebraical methods, he pointed out that the existence of even a tiny unexpected defect or irregularity in an apparently safe structure would be able to cause an increase of local stress which might be greater than the accepted breaking stress of the material and so might be expected to cause the structure to break prematurely.

In fact, using Inglis's formula (p. 67), it is easy to calculate that, if you were to scratch a girder of the Forth railway bridge, moderately hard, with an ordinary sharp pin, the resulting stress concentration should be sufficient to cause the bridge to break and fall into the sea. Not only do bridges seldom fall down when they are scratched with pins, but all practical structures such as machinery and ships and aeroplanes are infested with stress concentrations caused by holes and cracks and notches which, in real life, are only rarely dangerous. In fact they generally do no harm at all. Every now and then, however, the structure does break; in which case there may be a very serious accident.

When the implications of Inglis's sums began to dawn upon engineers some fifty or sixty years ago, they were apt to dismiss the whole problem by invoking the 'ductility' of the metals which they were accustomed to use. Most ductile metals have a stress-strain curve which is shaped something like figure 9, and it was commonly said that the overstressed metal at the tip of a crack simply flowed in a plastic sort of way and so relieved itself of any serious excess of stress. Thus, in effect, the sharp tip of the crack could be considered as 'rounded off' so that the stress concentration was reduced and safety was restored.

Like many official explanations, this one has the merit of being at least partly true, though in reality it is very far from being the whole story. In many cases the stress concentration is by no means fully relieved by the ductility of the metal, and the local stress does, in fact, quite often remain much higher than the commonly accepted 'breaking stress' of the material as determined from small specimens in the laboratory and incorporated in printed tables and reference books.

For many years, however, embarrassing speculations which were likely to undermine peoples' faith in the established methods of calculating the strength of structures were not encouraged. When I was a student Inglis's name was hardly ever mentioned and these doubts and difficulties were not much spoken about in polite engineering society. Pragmatically, this attitude could be partially justified, since, given a judiciously chosen factor of safety, the traditional approach to strength calculations – which virtually ignores stress concentrations – could generally be relied upon to predict the strength of most conventional metal structures. In fact it forms the basis of nearly all of the safety regulations which are imposed by governments and insurance companies today.

However, even in the best engineering circles, scandals occurred from time to time. In 1928, for instance, the White Star liner *Majestic* of 56,551 tons, which was then the largest and finest ship in the world, had an additional passenger lift installed. In the process rectangular holes, with sharp corners, were cut through several of the ship's strength decks. Somewhere between New York and Southampton, when the ship was carrying nearly 3,000 people, a crack started from one of these lift openings, ran to the rail, and proceeded down the side of the ship for many feet before it was stopped, fortuitously, by running into a port-hole. The liner reached Southampton safely and neither the passengers nor the press were told. By an extraordinary coincidence, very much the same thing happened to the second largest ship in the world, the American transatlantic liner *Leviathan*, at about the same time. Again the ship got safely into port and publicity was avoided. If the cracks had run a little further, so that these ships had actually broken in two at sea, the loss of life might have been severe.

Really spectacular accidents of this kind to large structures such as ships and bridges and oil-rigs became common only during and after the last war, and latterly they have been growing more, and not less, frequent. What has emerged rather painfully over a number of years – at a vast cost in life and property – is that, although the traditional view of elasticity as hammered out by Hooke and Young and Navier and by scores of nineteenth-century mathematicians is extremely useful and certainly ought not to be neglected or spurned, yet it is not really enough, by itself, to predict the failure of structures – especially large ones – with sufficient certainty.

The approach to structures through the concept of energy

> *I saw the different things you did,*
> *But always you yourself you hid.*
> *I felt you push, I heard you call,*
> *I could not see yourself at all.*

R. L. Stevenson, *A Child's Garden of Verses*

Until fairly recently elasticity was studied and taught in terms of stresses and strains and strength and stiffness, that is to say, essentially in terms of forces and distances. This is the way in which we have been considering it so far, and indeed I suppose that most of us find it easiest to think about the subject in this manner. However, the more one sees of Nature and technology, the more one comes to look at things in terms of energy. Such a way of thinking can be very revealing, and it is the basis of the modern approaches to the strength of materials and the behaviour of structures; that is, to the rather fashionable science of 'fracture mechanics'. This way of looking at things tells us a great deal, not only about why engineering structures break, but also about all sorts of other goings-on – in history and in biology, for instance.

It is a pity, therefore, that the whole idea of energy has been confused in many people's minds by the way in which the word is often used colloquially. Like 'stress' and 'strain', 'energy' is used to refer to a condition in human beings: in this case one which might be described as an officious tendency to rush about doing things and pestering other people. This use of the word has really only a tenuous connection with the precise, objective, physical quantity with which we are now concerned.

The scientific kind of energy with which we are dealing is officially defined as 'capacity for doing work', and it has the dimensions of 'force-multiplied-by-distance'. So, if you raise a weight of 10 pounds through a height of 5 feet, you will have to do 50 foot-pounds of work, as a result of which 50 foot-pounds of additional energy will be stored in the weight as what is called 'potential energy'. This potential energy is locked up, for the time being, in the system, but it can be released at will by allowing the weight to descend again. In doing so the released energy could be employed in performing 50 foot-pounds' worth of useful tasks, such as driving the mechanism of a clock or breaking the ice on a pond.

Energy can exist in a great variety of different forms – as potential energy, as heat energy, as chemical energy, as electrical energy and so on. In our material world, every single happening or event of whatever kind involves a conversion of energy from one into another of its many forms. In a physical sense that is what 'happenings' or 'events' are about. Such transformations of energy take place only according to certain closely defined rules, the chief of which is that you can't get something for nothing. Energy can neither be created nor destroyed, and so the total amount of energy which is present before and after any physical transaction will not be changed. This principle is called 'the conservation of energy'.

Thus energy may be regarded as the universal currency of the sciences, and we can often follow it through its various transformations by means of a sort of accounting procedure which can be highly informative. To do this, we need to use the right kind of units; and, rather predictably, the traditional units of energy are in a fine, rich state of muddle. Mechanical engineers have tended to use foot-pounds, physicists are addicted to ergs and electron-volts, chemists and dietitians like to use calories, but our gas bills come in therms and our electricity bills in kilowatt-hours. Naturally, all these are mutually convertible, but nowadays there is a good case for using the S I unit of energy, which is the Joule, that is the work done when one Newton acts through one metre.*

Although we can measure it in quite precise ways, many people find energy a more difficult idea to grasp than, say, force or distance. Like the wind in Stevenson's verse we can only apprehend it through its effects. Possibly for this reason the concept of energy came rather late into the scientific world, being introduced in its modern form by Thomas Young in 1807. The conservation of energy was not universally accepted until quite late in the nineteenth century, and it is really only since Einstein and the atom bomb that the enormous importance of energy as a unifying concept and as an underlying reality has been sufficiently appreciated.

There are, of course, a great many ways, chemical, electrical, thermal and so on, of storing energy until it is wanted. If we are going to use a mechanical means then we could use the method we have just been talking about, that is to say, the potential energy of a raised weight. However, this is rather a crude way of storing energy and, in practice, strain energy, the energy of a spring, is generally more useful and it has much more widespread applications in biology and engineering.

It is obvious that energy can be stored in a wound-up spring, but, as Hooke pointed out, official springs are only a special case of the behaviour of any solid when it is loaded. Thus every elastic material which is under stress contains strain energy, and it does not make much difference whether the stress is tensile or compressive.

If Hooke's law is obeyed, the stress in a material starts at zero and builds up to a maximum when the material is fully stretched. The strain energy per unit volume in the material will be the

---

*1 Joule (1J) = 10^7 ergs = 0·734 foot-pound = 0·239 calories. Note that one Joule is roughly the energy with which an ordinary apple would hit the floor if it fell off an ordinary table.

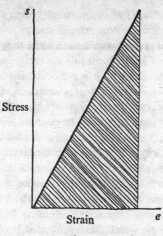

Figure 1. Strain energy = area under stress-strain curve = $\frac{1}{2}se$.

shaded area under the stress-strain diagram (Figure 1), which is

$$\frac{1}{2} \times \text{stress} \times \text{strain} = \frac{1}{2}se.$$

Cars, skiers and kangaroos

We are all of us familiar with strain energy in the springs of our car. In a vehicle with no springs there must be violent interchanges of potential and kinetic energy (energy of motion) every time a wheel passes over a bump. These energy changes are bad for the passengers and bad for the vehicle. Long ago some genius invented the spring, which is simply an energy reservoir which enables changes of potential energy to be stored temporarily as strain energy so as to smooth the ride and prevent the vehicle and its occupants from being racketed to bits.

Latterly engineers have spent a great deal of time and effort on the improvement of car suspensions, and no doubt they have been very clever about it. However, cars and lorries run on roads whose main purpose is, after all, to provide a smooth surface. The suspension of the car has only to even out the minor or residual bumps. The problem of designing a suspension for a car which had to be driven really fast across rough country would be a very difficult one. In order to store enough energy to cope with such a situation the steel springs would have to be very large and heavy and would in themselves constitute so much 'unsprung weight' that the whole project might prove to be impracticable.

Consider now the situation of a skier. In spite of the snow covering, most ski-runs are vastly more bumpy than any normal road. Even if a typical run could be covered with some effective non-skid surface, such as sand, so as to enable a car to go on it without slipping, any attempt to drive the car down the run at the speed of a fast skier – say 50 m.p.h. – would be suicidal, because the suspension would be completely inadequate to absorb the shocks. But, of course, this is exactly what the body of a skier has to do. In fact, much of this energy seems to be absorbed by the tendons in our legs, which, taken together probably weigh less than a pound.* Thus, if we are to ski fast without disaster or to perform other athletic feats, our tendons have to be able to store reliably and to give up again very large amounts of energy. This is partly what they are for.

Some approximate figures for the strain energy storage capabilities of various materials are given in Table 3. The relative efficiencies of natural materials and of metals may come as a surprise to engineers, and some light is thrown on the performance of skiers and other animals by the figures for tendon and steel. It will be seen that the strain energy storage per unit weight is about twenty times higher for tendon than it is for modern spring steels. Although, considered as devices for storing strain energy, skiers are more efficient than most machines, yet even a trained athlete cannot compete with a deer upon a hillside or a squirrel or a monkey in a tree. It might be interesting to know the percentage of the body weight given up to tendon in these animals, as compared with people.

Animals like kangaroos progress by bounding. At each landing,

---

*Since the oxygen consumption of the body is said to be higher during down-hill ski-ing than in any other human activity, much energy must also be got rid of in the muscles. However, most of the energy absorbed by the muscles is irrecoverable, and so the elastic strain energy storage of the tendons is no doubt to be preferred.

TABLE 3

Approximate strain energy storage capacities of various solids

Material	Working strain %	Working stress p.s.i.	Working stress MN/m²	Strain energy stored Joules × 10⁶ per cubic metre	Density kilograms per cubic metre	Energy stored Joules per kilogram
Ancient iron	0·03	10,000	70	0·01	7,800	1·3
Modern spring steel	0·3	100,000	700	1·0	7,800	130
Bronze	0·3	60,000	400	0·6	8,700	70
Yew wood	0·9	18,000	120	0·5	600	900
Tendon	8·0	10,000	70	2·8	1,100	2,500
Horn	4·0	13,000	90	1·8	1,200	1,500
Rubber	300	1,000	7	10·0	1,200	8,000

energy has to be stored in the creature's tendons, and I have been told by an Australian correspondent that the strain energy characteristics of kangaroo tendon are exceptionally good; but unfortunately I cannot quote any accurate figures. It occurs to me, however, that, if anyone should wish to revive the pogo-stick in a more efficient form, there would be a good deal to be said for making the spring from kangaroo tendon, or indeed from any form of tendon. Light aircraft, which have to be designed for bad landings on rough ground, often have their undercarriages sprung by means of rubber cords which have a strain energy capacity much better than that of steel springs, and are also better than tendon, though they are less durable.

Besides its role in the suspensions of cars and aeroplanes and animals, strain energy plays an even more important part in the strength and fracture of all kinds of structures. However, before we pass on to the subject of fracture mechanics it may be worth spending a little time in discussing yet another application of strain energy, that is in weapons such as bows and catapults.

Bows

I will bring you the great bow of the divine Odysseus, and whosoever shall most easily string the bow with his hands, and shoot through all the twelve axes, with him will I go and forsake this house, this house of my marriage, so beautiful and filled with fair things, which I think I shall yet remember, aye, in a dream.

Penelope, in Homer, *Odyssey* XXI

The bow is one of the most effective ways of storing the energy of human muscles and releasing it to propel a missile weapon. The English longbow, which did so much execution at Crecy (1346) and Agincourt (1415), was nearly always made from yew. Because yew timber has not much commercial value nowadays, little scientific work was done on it until recently. However, my colleague Dr Henry Blyth, who is doing research on ancient weapons, has ascertained that yew (*Taxus baccata*) has a fine-scale morphology which is rather different from other timbers and seems to be peculiarly adapted for storing strain energy. Thus yew probably really is better than other woods for making bows.

Contrary to popular belief, English longbows were not, as a rule, made from English yew-trees, whether grown in church-yards or elsewhere. Most English bows were made from Spanish yew and it was legally compulsory to import consignments of Spanish bow-staves with each shipment of Spanish wine. In fact the yew-tree grows well, not only in Spain, but all over the Mediterranean area. It is growing wild today among the ruins of Pompeii for instance. In spite of this, one seldom hears of the use of yew bows in Spain or in the Mediterranean countries, either during the Middle Ages or in antiquity. Their use was almost confined to England and France and, to some extent, Germany and the Low Countries. English depredations generally stopped somewhere around Burgundy and hardly ever spread south of the Alps or the Pyrenees.

Although these facts seem surprising at first sight, Henry Blyth points out that, because of its rather special constitution, the mechanical properties of yew deteriorate more rapidly with in-creasing temperature than do those of other timbers. A yew bow cannot be used reliably above 35° C. As a weapon it is therefore pretty well confined to cool climates and is unsuitable for use in the Mediterranean summer. Thus, although yew wood was used for arrows, it was seldom used for bows in Mediterranean countries.

For this reason what was called a 'composite' bow was developed in these countries. Such bows had a core of wood which, being near the middle of the thickness of the bow, was only lightly stressed. To this core was glued a tension surface made from dried tendon and a compression face made from horn. Both these materials are even better at storing energy than yew. Furthermore they retain their mechanical properties better than yew in hot weather. After all, an animal normally operates at about 37° C. In practice, tendon does not deteriorate appreciably below about 55° C. As against this, dried tendon slackens and behaves badly in damp weather.

Composite bows of this kind were used both in Turkey and elsewhere down to comparatively recent times. Lord Aberdeen (1784–1860), travelling to the Congress of Vienna in 1813, wrote of the use of Tartar troops, armed with what seem to have been

composite bows, against the armies of Napoleon which were retreating through eastern Europe. There is a good deal of evidence
that composite bows were better in many respects than the English
longbow. However, whereas the longbow was essentially a cheap
and simple weapon to manufacture, the composite bow was a
much more sophisticated affair, and presumably expensive. Greek
bows were composite bows, and the bow of Odysseus, like that of
Philoctetes, seems to have been a pretty special job.

Which brings us back to the unfortunate Penelope and the task
she set her suitors of stringing the bow of Odysseus. As we all
know, this turned out to be beyond the strength of any of them,
even the technically-minded Eurymachus: 'And now Eurymachus
was handling the bow, warming it on this side and on that before
the heat of the fire; yet even so he could not string it, and in his
great heart he groaned mightily.' But after all, why bother? Why
didn't the suitors, or Odysseus, or anybody else, just use a longer
string?

The answer to this is 'for a very good scientific reason' – which
is as follows. The energy which a man can put into a bow is
limited by the characteristics of the human body. In practice, one
can draw an arrow back about 0·6 metres (24 inches), and even a
strong man cannot pull on the string with a force of more than
about 350 Newtons (80 lb.). It follows that the available muscular
energy must be around 0·6 metre × 350 Newtons, or about 210
Joules. This is the most that is available, and we want to store as
much of it as possible as strain energy in the bow.

If we suppose that the bow is initially virtually unstressed and
that the string is almost slack to begin with, then the archer
starts to draw his arrow with a pull which is initially nearly zero,
and he only works up to his greatest pull when the string reaches
its maximum extension. This is expressed diagrammatically in
Figure 2. In such a case, the energy put into the bow is the area of
the triangle A B C, which cannot be more than *half* of the available
energy, i.e. 105 Joules.

In practice the measured energy which was stored in an English
longbow was a little less than this figure. However, Homer specifically says that the bow of Odysseus was *palintonos*, that is, 'bent
or stretched backwards'. In other words the bow was initially

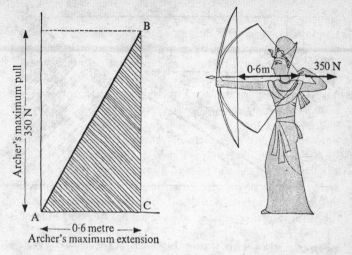

Figure 2. Energy stored in bow = $\frac{1}{2} \times 0.6 \times 350 = 105$ Joules.*

bent in the opposite or 'wrong' direction, so that considerable force had to be applied to it before it could be strung.

When the bow is strung in this way the archer is no longer

Figure 3. Greek stringing bow (vase painting).

starting to draw the bow from an initial condition of zero stress and strain; and, by intelligent design, it is now possible to arrange for the force-extension diagram to look something like Figure 4.

* Figures 2 and 4 are, of course, schematic. Generally the force-draw diagram will not be a straight line; but the same principle applies.

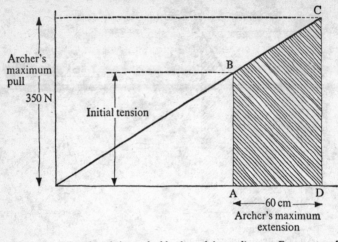

Figure 4. Why a bow is 'stretched backwards' or *palintonos*. Energy stored in bow is now area ABCD ≈ 170 Joules.

The area ABCD under such a diagram is now a very much higher fraction of the total available energy and might perhaps reach about 80 per cent of it. So it is possible that about 170 Joules of energy can now be stored in the bow, as against only about 105 Joules for the bow that is not *palintonos*. This is clearly a great improvement for the archer – quite apart from any advantage it might have had for Penelope.

In fact all bows are more or less pre-stressed, in the sense that some kind of effort is needed to string them. However, since the longbow is a 'self-bow', that is to say, it is made from a stave which has been split from a log of timber and is therefore initially nearly straight, the effect in this case was small. It is much easier to arrange for the best initial shape with a composite bow, and these had usually a very characteristic form, from which we get the shape of a 'Cupid's bow' (Figure 5).

Because the strain energy storage of horn and tendon, as materials, is better than that of yew, a composite bow can be made shorter and lighter than a wooden one. This is why we talk of a wooden bow as a 'long' bow. The composite bow could be made small enough to be used on horseback, as was indeed done by the

Parthians and the Tartars. The Parthian bow was handy enough for the cavalrymen to be able to shoot backwards, as they retreated, at their Roman pursuers; from this we get the phrase 'a Parthian shot'.

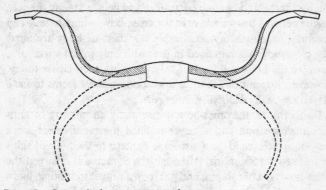

Figure 5.   Composite bow, unstrung and strung.

Catapults

The greatest period of classical Greece came to an end when Athens fell in 404 B.C., and during the fourth century the Greek democratic governments declined and were superseded by dictatorships or 'tyrannies', which may have been more effective militarily, politically and economically. Both ashore and afloat the technology of warfare was changing, and the new rulers considered that there was a need for more modern and more mechanized weapons. Moreover, as the absolute masters of increasingly prosperous states, the dictators could well afford to pay the bills.

Development began in Greek Sicily. Dionysius I was a remarkable man who had risen from being a petty clerk in a government office to become Tyrant of Syracuse. During most of his reign, which lasted from 405 to 367 B.C., he made his country the leading power in Europe. As a part of his military programme he founded what was probably the first government research laboratory for weaponry, and for this establishment he recruited the best mathematicians and the best craftsmen from all over the Greek world.

The natural starting point for Dionysius's experts was the traditional composite hand-bow. If one mounts such a bow upon some kind of stock and arranges to draw the string by means of mechanical gearing or levers, then the bow itself can be made much stiffer and so be enabled to store and deliver several times as much energy. Thus we arrive at the cross-bow, whose missile can generally penetrate any practicable thickness of body-armour.* The cross-bow has remained in use, with only minor variations, down to the present time. It is said to be in use in Ulster today. However, it is curious that, as a weapon, it never seems to have played any really decisive military role.

Furthermore, the cross-bow is essentially an infantry or anti-personnel weapon and it never fulfilled the requirement for a weapon which could do worthwhile damage to the hulls of ships or to fixed fortifications. Although the Syracusans enlarged the cross-bow type of catapult and put it on a proper mounting, like a gun-mounting, there seem to be certain physical limitations to this line of development, and catapults of the bow type do not seem ever to have been powerful enough to breach the heavy masonry of fortresses.†

The next step was therefore to abandon the bow type of construction and to store the strain energy in twisted skeins of tendon,‡ much like the skeins of rubber cord which are used to drive model aeroplanes. In such a skein all the cords, that is, the whole of the tendon material, are being stretched in tension as the skein is twisted, so that as an energy storage device it is very effective indeed.

There are various ways in which skeins of tendon rope can be

---

* On the other hand the rate of shooting of a cross-bow cannot match that of a hand-bow. The English longbow, for instance, could discharge up to fourteen arrows a minute and thus, when used *en masse*, could put up a very formidable cloud or barrage of missiles. It is calculated that about *six million* arrows were shot at Agincourt.

† Recent finds at Kouklia in Cyprus indicate the existence of military catapults during the fifth century, though nothing is known about them. In any case Dionysius's seems to have been the first 'scientific' approach to the problem.

‡ These were probably derived from the 'Spanish windlass' used in ancient ships. See Chapter 11, p. 224.

used in weaponry, but by far the best was the device known to the Greeks as the *palintonon* and to the Romans as the *ballista*. In this very lethal piece of artillery there were two vertical tendon springs, each of them twisted by means of a rigid arm or lever, something like a capstan bar (Figure 6). The ends of these two arms were joined by a heavy bowstring, and the whole device worked much after the fashion of a bow. Indeed it got its Greek

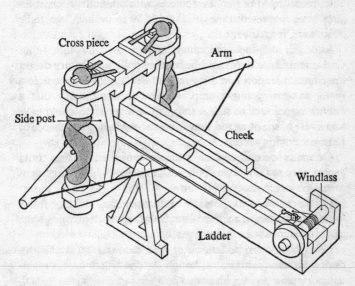

Figure 6.   A sketch of what original Greek catapults may have looked like.

name from the fact that, in their relaxed position, the two arms point forward, like the arms of a composite bow; and the catapult was strung (by means of a powerful winch) in much the same way as a bow. The missile, which was often a stone ball, was propelled down a track which also served to mount the windlass that was needed to operate the weapon, whose draw force might be as high as a hundred tons.

The Romans copied the Greek catapults and Vitruvius, who was an artillery officer under Julius Caesar, has left us a handbook on *ballistae* which makes interesting reading. These weapons were made in sizes which ranged from one throwing a 5 lb. (2 kg) missile to one throwing a 360 lb. (150 kg) one. The effective range of all sizes was about a quarter of a mile or 400 metres. The standard Roman siege *ballista* seems to have been one throwing a 90 lb. (40 kg) ball.

At the final, dramatic, siege of Carthage in 146 B.C. the Romans filled in part of the shallow lagoon which lies against the city wall and proceeded to breach the defences with catapults. Archaeologists have recovered no fewer than 6,000 stone balls, weighing 90 lb. each, from the site.

Although ship-mounted catapults were used by both Julius Caesar and Claudius to clear the beaches of ancient Britons during their assault-landings on this island, the catapult never became a really dangerous ship-to-ship weapon. It seems likely that a *ballista* big enough to sink a ship with a single shot would have had a rate of shooting too slow for it to have had much chance of hitting a moving vessel.

Catapults sometimes threw incendiary missiles, but fires could generally be put out quite easily in simple ships which were full of men. One ingenious admiral won a sea-battle in 184 B.C. by shooting at the enemy brittle pots filled with poisonous snakes, but this lead does not seem to have been followed up. On the whole, catapults were not a success at sea.

Nevertheless, the *palintonon* or *ballista* was a most effective device for land warfare, although its construction and maintenance were a very sophisticated business indeed, and the Roman artillery officers and N.C.O.s must have been highly competent people. With the passing of the Roman Empire and of Roman technology such weapons became impracticable and were forgotten.* Medieval siege-warfare was reduced to using the weight-catapult or 'trebuchet'.

This was a pendulum-like device using the potential energy of a

* During the 1940 invasion scare two versions of the Roman *ballista* were n.ade for use by the Home Guard in England. These weapons were intended to project petrol bombs against German tanks. However, since the range of both of these catapults was only about a quarter of that of the classical prototypes, it seems likely that their designers had omitted to read Vitruvius sufficiently carefully.

raised weight. Even a large trebuchet was unlikely to involve raising more than say a ton (10,000 Newtons) through about 10 feet (3 metres). Thus the greatest potential energy stored probably did not much exceed 30,000 Joules. The same amount of strain energy could be stored in ten or twelve kilograms of tendon. Thus even a big trebuchet probably disposed of only about a tenth of the energy of the *palintonon*. Furthermore the efficiency of energy

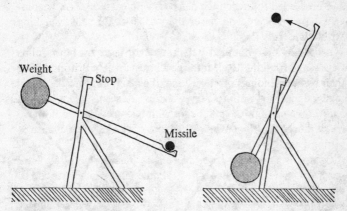

Figure 7. The trebuchet or medieval weight-catapult – a most inefficient contrivance.

conversion seems to have been much lower. At its best the trebuchet could probably only make a nuisance of itself by lobbing big stones over a fortress wall; any assault upon heavy masonry would have been ineffectual.

Regarded as machines for the conversion of energy, the bow and the *palintonon* both work on similar principles; it is not generally realized just how efficient an energy exchange mechanism is involved. In crude machines like the trebuchet, most of the energy which was available when the weapon was discharged went into accelerating the heavy lever or throwing-arm of the device and was ultimately lost in the necessary stop or braking system.

With a bow or a *palintonon*, when the bowstring is first released, some of the stored strain energy is communicated directly to the missile as kinetic energy. More of the available energy, however, is being used to accelerate the arms of the bow or the catapult, where it is temporarily stored as kinetic energy, much as it is in the trebuchet. In this case, though, as the discharge mechanism proceeds, the moving arms are slowed down, not by a fixed stop, but by the bowstring itself as it straightens and tautens. This further increases the tension in the string, enabling it to push yet

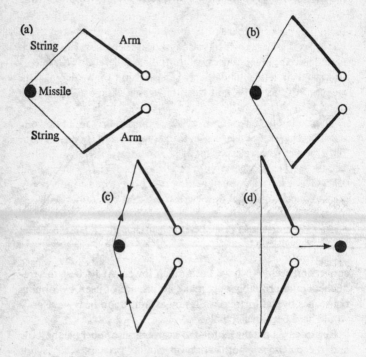

Figure 8. Diagram of the mechanism of the *palintonon* or *ballista*.
    (a) Ready to shoot. All the energy is stored in the tendon springs.
    (b) Early stage of shooting operation. Heavy arms are being *accelerated* and so pick up much of the energy from the springs.
    (c) Late stage of shooting operation. Heavy arms are being decelerated by increased tension in the string, and so their kinetic energy is transferred to the missile.
    (d) Missile on its way, containing virtually all the energy which was stored in the system.

harder on the missile and so speed it on its way. Thus, much of the kinetic energy stored in the arms is recovered.

The mathematics of bows and catapults is difficult and, even when one has written down the equations of motion, they cannot be solved analytically. Fortunately, however, another colleague of mine, Dr Tony Pretlove, has been sufficiently interested in the problem to put the whole thing on a computer. It transpires that, rather surprisingly, the energy transfer process is in theory virtually 100 per cent efficient. In other words, practically the whole of the strain energy which was stored in the device can be converted into the kinetic energy of the missile. Therefore little energy is wasted or left behind to provide a recoil or to damage the weapon. In this respect, at least, bows and catapults are a great improvement on guns.

One consequence of these facts is, I think, fairly well known to most archers, at least in a practical sort of way. This is that one must never, never, never 'shoot' a bow or a catapult without a proper arrow or other appropriate missile. If this is attempted, then there is no safe way of getting rid of the stored strain energy, and, not only may the bow be broken, but the archer will very possibly be hurt as well.

Resilience or bounciness

> *A wet sheet and a flowing sea,*
> *A wind that follows fast*
> *And fills the white and rustling sail*
> *And bends the gallant mast.*

Allan Cunningham, *A Wet Sheet and a Flowing Sea*

When Galileo settled down at Arcetri in 1633 to work on elasticity, one of the first questions he asked himself was 'What are the factors which affect the strength of a rope or a rod when it is pulled? Does the strength depend, for instance, upon the length of the rope?' Elementary experiments showed that the force or weight which is needed to break a uniform rope by pulling on it steadily is unaffected by how long it is. This result is what we should expect from common sense, but the news has been some time in getting around and one still meets quite a number of people who are convinced that a long piece of string is 'stronger' than a short one.

Of course these people are not just being silly, for it all depends on what you mean by 'stronger'. The steady force or pull required to break a long string will indeed be the same as that needed to break a short one, *but* the long string will stretch further before it breaks and it will therefore require more *energy* to break it, even though the force which is applied and the stress which is in the material remain the same. Put in a slightly different way, a long string will cushion a sudden blow by stretching elastically under the load, so that the transient forces and stresses which result are reduced. In other words it acts rather like the suspension of a car.

Thus in a situation where the load is jerky a long string may well be effectively 'stronger' than a short one. This is why the bodies of eighteenth-century carriages were frequently slung from the chassis by means of very long leather straps, which were better able than short ones to resist the jolts imposed by eighteenth-century roads. Again, anchor cables and tow ropes generally break, not from a steady load, but from sudden jerks, and so it is generally better to arrange for them to be as long as possible. Those who are liable to encounter large dry-docks or oil-rigs under tow at sea at night or in thick weather do well to bear in mind that each of the tugs is probably towing by means of nearly a mile of steel wire. These nautical processions therefore cover an enormous area of sea and can be terrifying to the casual seafarer.*

This quality of being able to store strain energy and deflect elastically under a load without breaking is called 'resilience', and it is a very valuable characteristic in a structure. Resilience may be defined as 'the amount of strain energy which can be stored in a structure without causing permanent damage to it'.

---

* Actually, much of the resilience of anchor cables and tow ropes comes from their own weight, which causes them to sag. This is one of the reasons for preferring heavy wire or chain to organic ropes, which are much lighter.

Of course, in order to get resilience, it is not necessary to use a very long rope, such as a wire cable. It is often convenient to use much shorter members, such as the helical springs which are used in the buffers of railway trains, or pads of soft material such as are used for ships' fenders, or materials of low Young's modulus, like the foamed rubbers or plastics which are often used for packaging delicate apparatus. Such things are frequently able to stretch or contract much more in relation to their length and so store more strain energy per unit volume. The excellence of the suspensions of skiers and animals is due, in part, to the comparatively low moduli and high extensions of tendon and other tissues.

On the other hand, although low stiffness and high extensibility promote energy absorption, and so make it more difficult to break a structure by means of a blow, it is only too easy to make a structure which is too floppy for its purpose. This usually limits the amount of resilience which can be designed into a structure. Things like aeroplanes and buildings and tools and weapons have to be pretty rigid in order to do their job. In this respect most structures have to be a compromise between stiffness and strength and resilience, and the achievement of the best compromise is likely to tax the skill of a designer severely.

The optimum condition may vary, not only between different types and classes of structures, but also between different parts of the same structure. In this respect Nature is at an advantage since she has at her disposal an enormous range of elastic properties in the different biological tissues. A simple but interesting example occurs in an ordinary spider's web. The web is subject to impact loads arising from flies blundering into it, and the energy of these blows must be absorbed by the resilience of the threads. It turns out that the long radial threads, which form the main load-carrying part of the structure, are three times as stiff as the shorter circumferential threads which have the duty of actually catching the flies.

Naturally, there are many other ways of storing strain energy and getting resilience than by using tension members, such as ropes or spider's threads, or compression members, such as railway buffers and ships' fenders. Any shape of structure which is capable of being deflected elastically will have much the same effect. Probably the commonest arrangement is to absorb energy by bending, like bows and gallant masts. This is what happens in plants and trees and in most car springs. High-quality swords are expected to be able to recover, elastically, after they have been bent so that the tip touches the hilt.

Strain energy as the cause of tensile fracture

> Starting aside like a broken bow.

Psalm 78

A reasonable amount of resilience is an essential quality in any structure; otherwise it would be unable to absorb the energy of a blow. Up to a point, the more resilient a structure is the better. Such highly sophisticated devices as the Viking ships and the American horse buggy were very flexible and resilient indeed. As long as they are not grossly overloaded, such structures will recover when the load is taken off and all will be well. However, if we do overload them, then of course sooner or later they will break.

Now to break any material in tension a crack must spread right across it. However, to create a new crack requires a supply of energy – as we shall shortly see – and this energy has to come from somewhere. As we have said, it is quite possible to break a bow by 'shooting' it without an arrow. What happens is that the strain energy which was stored in the bow can no longer be disposed of safely as kinetic energy in the arrow, and so some of it is employed in producing cracks within the material of the bow itself. In other words the bow has used its own strain energy to destroy itself. The broken bow is, however, only a special case of *all* kinds of fracture.

All elastic substances which are under load contain greater or less amounts of strain energy, and this strain energy is always potentially available for the self-destructive process which we call 'fracture'. In other words the stored-up strain energy or resilience may be used to pay the energy-price of propagating a crack through the structure and so causing it to break. In a resilient structure there may be a lot of strain energy around, and the same sort of energy which the Romans used to batter down the massive walls of Carthage can equally well be employed to enable a super-tanker to break herself into two halves.

According to the modern view of the subject, when we break a structure by loading it in tension, we ought not to regard fracture as being caused *directly* by the action of the applied load pulling on the chemical bonds between the atoms in the material. That is to say, it is not the consequence of the simple action of a tensile stress as the classical text-books would have us believe.* The direct result of increasing the load on the structure is only to cause more strain energy to be stored within its material. The sixty-four thousand dollar question whether the structure actually breaks at any particular juncture depends upon whether or not it is possible for this strain energy to be converted into fracture energy so as to create a new crack.

Modern fracture mechanics is therefore less concerned with forces and stresses than with how, why, where and when strain energy can be turned into fracture energy. Of course, in simple cases like ropes and rods the classical concept of a critical breaking stress is usually an adequate guide, but in large or complicated structures, such as bridges or ships or pressure vessels, it has proved to be a dangerous oversimplification, as we have seen. What comes out of recent theory is that, whether a structure is subjected to a sudden blow or to a steady load, tensile fracture depends *chiefly* upon:

1. The price in terms of energy which has to be paid in order to create a new crack.

2. The amount of strain energy which is likely to become available to pay this price.

3. The size and shape of the worst hole or crack or defect in the structure.

The fact that the amount of energy required to break any given cross-section of material varies very greatly indeed between different solids is easily confirmed, for instance by hitting first a glass

* The 'true' or theoretical maximum tensile stress required actually to pull the atoms apart is very high indeed, far higher than the 'practical' strength determined by means of ordinary tensile tests. See *The New Science of Strong Materials*, Chapter 3.

jar and then a tin-can with a hammer. The quantity of energy required to break a given cross-section of a material defines its 'toughness', which is nowadays more often called its 'fracture energy' or 'work of fracture'. This property is quite different to and separate from the 'tensile strength' of the material, which is defined as the *stress* (not the energy) needed to break the solid. The toughness or work of fracture of a material has a very important effect upon the practical strength of a structure – especially a large one. For this reason we must spend a little time in talking about the work of fracture of various kinds of solids.

Fracture energy or 'work of fracture'

Since, when a solid is broken in tension, at least one crack must be made to spread right across the material, so as to divide it into two parts, at least two new surfaces will have to be created which did not exist before fracture. In order to tear the material apart in this way and produce these new surfaces it is necessary to have broken all the chemical bonds which previously held the two surfaces together.

The quantity of energy which is needed to break most kinds of chemical bonds is well known – at least to chemists – and it turns out that, for most of the structural solids with which we are concerned in technology, the total energy needed to break all the bonds on any one plane or cross-section* is very much the same and does not differ widely from 1 Joule per square metre.

When we are dealing with the range of materials which are, rather understandably, called 'brittle solids' – which includes stone and brick and glass and pottery – this is nearly all the energy we have to provide in order to cause fracture. As a matter of fact, 1 J/m² is really rather a pathetically small amount of energy. It is a sobering thought that, on the simplest theory, the strain energy which could be stored in one kilogram of tendon would 'pay' for the production of 2,500 square metres (over half

*This is often the same thing as the 'free surface energy', which is closely related to the surface tension of both liquids and solids, and which is frequently bandied about in discussions on materials science. See for instance, *The New Science of Strong Materials*, Chapter 3.

an acre) of broken glass surface – which accounts for the effects of bulls in china-shops. This is why a bricklayer can break a brick neatly in half with a light tap from his trowel and it is why we have only to be a little clumsy in order to break a plate or a tumbler.

Naturally, this is the reason why, if we can possibly avoid it, we do not use 'brittle solids' in applications where they are in tension. These materials are brittle not, primarily, because they have low tensile strengths – that is to say they need a low *force* to break them – but rather because it needs only a low *energy* to break them.

The technical and biological materials which are actually used in tension, and used with comparative safety, all require a great deal more energy in order to produce a new fracture surface. In other words, the 'work of fracture' is very much higher – enormously higher – than is the case with brittle solids. For a practical

TABLE 4

Very approximate figures for the work of fracture and tensile strengths of some common solids

Material	*Approximate* work of fracture J/m^2	Approximate tensile strength (nominal) MN/m^2
Glass, pottery	1–10	170
Cement, brick, stone	3–40	4
Polyester and epoxy resins	100	50
Nylon, polythene	1,000	150–600
Bones, teeth	1,000	200
Wood	10,000	100
Mild steel	100,000–1,000,000	400
High tensile steel	10,000	1,000

tough material the work of fracture usually lies between 10^3 J/m^2 and 10^6 J/m^2. *Thus the energy which is needed to cause fracture in wrought iron or mild steel may be about a million times as high as that needed to break the equivalent cross-section of glass or pottery, although the static tensile strengths of these materials are not very different.* This is why a table of 'tensile strengths', such as Table 2, (p. 56), can be a highly misleading document when it comes to the choice of a material for a particular service. It is also why the classical theory of elasticity, based mainly on forces and stresses, which has been laboriously evolved over hundreds of years – and still more laboriously taught to students – is really inadequate, by itself, to predict the behaviour of real materials and structures.

Although the detailed mechanisms whereby such enormous amounts of energy can be absorbed within tough materials as 'work of fracture' are often subtle and complicated, the broad principle is really very simple. In a 'brittle' solid the work done during fracture is virtually confined to that which is needed to break the chemical bonds at, or very near to, the new fracture surface. As we have seen, this energy is small and amounts only to

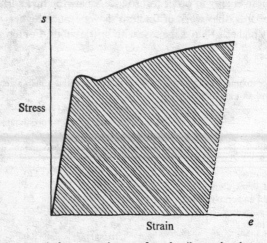

Figure 9.   A typical stress-strain curve for a ductile metal such as mild steel. The shaded area is related to the work of fracture of the metal.

about 1 J/m^2. In a tough material, although the strength and the energy of any individual bond remains the same, the fine structure of the material is disturbed to a very much greater depth during the breaking process. In fact it may be disturbed to a depth of well over a centimetre: that is, to a depth of about 50 million atoms below the visible fracture surface. Thus if only one in fifty of these atomic bonds is broken during the process of disturbance then the work of fracture – the energy needed to produce the new surface – will be increased a millionfold, which, as we have seen, is about what really does happen. In this way molecules living deep within the interior of the material are able to absorb energy and to play their part in resisting fracture.

The high work of fracture of the soft metals is primarily due to the fact that these materials are 'ductile'. This means that, when they are pulled in tension, the stress-strain curve departs from Hooke's law at quite a moderate stress, after which the metal deforms plastically, rather like plasticine (Figure 9). When a rod or sheet of such a metal is broken in tension the material is pulled out before it breaks after the fashion of treacle or chewing gum; the broken ends will then be tapered or conical and will look rather like Figure 10. This form of fracture is often called 'necking'.

Thick metal plate

Thin metal sheet

Figure 10. The work of fracture is proportional to the volume of metal plastically distorted, i.e. to the shaded area, and thus is roughly as t^2. Hence the work of fracture of thin sheet may be very low.

Necking and similar forms of ductile fracture can take place because a great many of the innumerable layers of atoms in the metal crystals are enabled to slide over each other by means of what is called the 'dislocation mechanism'. Dislocations not only enable layers of atoms to slide over each other like a pack of cards but they also absorb energy – quite a lot of energy. The result of all this slipping and sliding and stretching in the crystals is that the metal is enabled to distort and a great deal of energy is got rid of.

The dislocation mechanism,* which was originally postulated by Sir Geoffrey Taylor in 1934, has been the subject of intensive academic research over the last thirty years. It turns out to be an extraordinarily subtle and complicated affair. What takes place inside so apparently simple a thing as a piece of metal seems to be quite as clever as many of the mechanisms in living biological tissues. Yet the funny thing is that this clever mechanism cannot possibly be purposive, if only because Nature has nothing, so to speak, to gain from it, since she never makes any structural use of metals, which very seldom occur native in the metallic state anyway. However this may be, dislocations in metals have been of enormous benefit to engineers and might almost have been invented for their benefit, since they not only result in metals being tough but also enable them to be forged and worked and hardened.

Artificial plastics and fibrous composites have other work of fracture mechanisms which are quite different from those in metals but which are fairly effective. Biological materials seem to have developed methods of achieving high works of fracture which are very cunning indeed. That in timber, for instance, is exceptionally efficient, and the work of fracture of wood is, weight for weight, better than that of most steels.†

Let us now go on to discuss how the strain energy in a resilient structure manages to get turned into work of fracture. If you like, what is the *real* reason why things break?

Griffith – or how to live with cracks and stress concentrations

Ony rollin's better than pitchin' wi' superfeecial cracks in the tail-shaft.

Rudyard Kipling, *Bread upon the Waters* (1895)

As we said at the beginning of this chapter, all technological structures contain cracks and scratches and holes and other de-

---

* See *The New Science of Strong Materials*, Chapters 3 and 9, for an elementary account of the dislocation mechanism; for a fuller description see, for instance, *The Mechanical Properties of Matter* by Sir Alan Cottrell (John Wiley, 1964 etc.).

† Again, see *The New Science of Strong Materials* (second edition), Chapter 8.

fects; ships and bridges and aircraft wings are liable to all sorts of accidental dents and abrasions and we have to learn to live with them as safely as may be, in spite of the fact that, according to Inglis, the local stress at many of these defects may be well above the official breaking stress of the material.

How and why we are generally able to live with these high stresses without catastrophe was propounded by A. A. Griffith (1893–1963) in a paper which he published in 1920, just twenty-five years after Kipling's splendid story about a crack. Since Griffith was only a young man in 1920, practically nobody paid any attention. In any case Griffith's approach to the whole problem of fracture by way of energy, rather than force and stress, was not only new at the time but was quite foreign to the climate of engineering thinking, then and for many years afterwards. Even nowadays too many engineers do not really understand what Griffith's theory is all about.

What Griffith was saying was this. Looked at from the energy point of view, Inglis's stress concentration is simply a *mechanism* (like a zip-fastener) for converting strain energy into fracture energy, just as an electric motor is simply a mechanism for converting electrical energy into mechanical work or a tin-opener is simply a mechanism for using muscular energy to cut through a tin. None of these mechanisms will work unless it is continually supplied with energy of the right sort. The stress concentration is quite good at its job but, if it is to keep on prising the atoms of a material apart, then it needs to be kept fed with strain energy. If the supply of strain energy dries up, the fracture process stops.

Now consider a piece of elastic material which is stretched and then clamped at both ends so that, for the present, no mechanical energy can get in or out. Thus we have a closed system containing just so much strain energy.

If a crack is to propagate through this stretched material then the necessary work of fracture will have to be paid for in energy and the terms are strictly cash. If, for convenience, we consider our specimen to be a plate of material one unit thick, then the energy bill will be WL where W = work of fracture and L = length of crack. Note that this is an energy debt, an item on the debit side of the energy account, although as a matter of fact no credit is given. This debit increases linearly or as the first power of the length, L, of the crack.

This energy has to be found immediately from internal resources, and since we are dealing with a closed system, it can only come from some relaxation of strain energy within the system. In other words, somewhere in the specimen the stress must be diminished.

This can occur because the crack will gape a little under stress and thus the material immediately behind the crack surfaces is relaxed (Figure 11). Roughly speaking, two triangular areas –

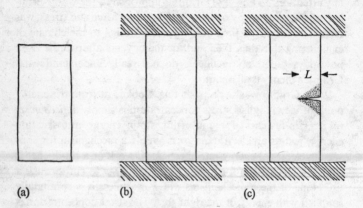

(a)   (b)   (c)

Figure 11. (a) Unstrained material.
           (b) Material strained and rigidly clamped. No energy can get in or out of the system.
           (c) Clamped material is now cracked. Dotted areas relax and give up strain energy, which is now available to propagate the crack still further.

which are shaded in the diagram – will give up strain energy. As one might expect, whatever the length, L, of the crack, these triangles will keep roughly the same proportions, and so their areas will increase as the *square* of the crack length, i.e. as L^2. Thus the strain energy release will increase as L^2.

Thus the core of the whole Griffith principle is that, while the energy *debt* of the crack increases only as L, its energy *credit* increases as L^2. The consequences of this are shown graphically in

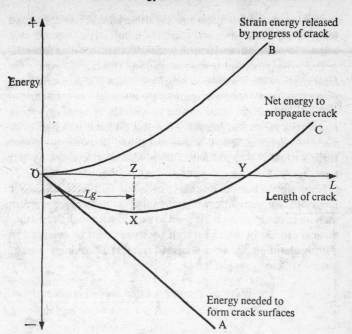

Figure 12. Griffith energy release, or why things go pop.

Figure 12. O A represents the increased energy requirement as the crack extends, and it is a straight line. OB represents the energy released as the crack propagates, and it is a parabola. The *net energy balance* is the sum of these two effects and is represented by OC.

Up to the point X the whole system is consuming energy; beyond point X energy begins to be released. It follows that there is a critical crack length, which we might refer to as L_g, which is called the 'critical Griffith crack length'. Cracks shorter than this are safe and stable and will not normally extend; cracks longer than L_g are self-propagating and very dangerous.* Such cracks

*It might perhaps be supposed that L_g would correspond to O Y on the diagram, but a little thought will show that this is not the case. The negative amount of energy, ZX, which we have to feed into the system to get the crack going represents the margin of safety or threshold energy. (This is, in fact, the true 'factor of safety'.)

spread faster and faster through the material and inevitably lead to an 'explosive', noisy and alarming failure. The structure will end with a bang, not a whimper, and very possibly with a funeral.

The most important consequence of all this is that, *even if the local stress at the crack tip is very high – even if it is much higher than the 'official' tensile strength of the material – the structure is still safe and will not break so long as no crack or other opening is longer than the critical length L_g.* It is this principle which gives us our main defence against undue alarm and despondency about Inglis's stress concentrations. This is why holes and cracks and scratches are not even more dangerous than they are.

Naturally we want to be able to calculate L_g numerically. As it turns out, for straightforward conditions, this is much simpler than we have any reasonable right to expect. Although the mathematical process by which Griffith got there might be regarded as slightly alarming, the actual finished result is disarmingly simple, indeed brilliantly simple, for

$$L_g = \frac{1}{\pi} \times \frac{\text{Work of fracture per unit area of crack surface}}{\text{Strain energy stored per unit volume of material}}$$

or, put algebraically,

$$L_g = \frac{2WE†}{\pi s^2}$$

where W = work of fracture in J/m^2 for each surface

E = Young's modulus in Newtons/m^2

s = average tensile stress in material near the crack (taking no account of stress concentrations) in Newtons/m^2

L_g = critical crack length in metres.

(Caution, Newtons, not Meganewtons.)

So the length of a safe crack depends simply upon the ratio of the value of the work of fracture to that of the strain energy stored in the material – in other words it might be considered as inversely proportional to the 'resilience'. In general, the higher the resilience

† Because strain energy = $\frac{1}{2} es$, which can be written $s^2/2E$, since $E = s/e$.

the shorter the crack one can afford to put up with. This is another example of not being able to have things both ways.

As we have seen, rubber will store a great deal of strain energy. However, its work of fracture is quite low and so the critical crack length, L_g, for stretched rubber is quite short, usually a fraction of a millimetre. This is why, when we stick a pin into a blown-up balloon, it bursts with a very satisfactory pop. Thus, although rubber is highly resilient and will stretch a long way before it breaks, when it does break, it breaks in a brittle manner, very much like glass.

One solution to the problem of how to be resilient and also tough is that provided by cloth and basket-work and wooden ships and horse-drawn vehicles. In these things the joints are more or less loose and flexible and so energy is absorbed in friction – that is what all the squeaks are about. However, although hedges and birds' nests are pretty resistant to attack, this way of going about things is not often used by modern engineers, except perhaps in the case of car tyres, where the rubber is saved from being unduly brittle by the incorporation of canvas and cords.

It will be seen that L_g shortens very rapidly as the stress, s, increases. Thus, if we want to be able to accommodate a good long crack safely at a reasonably high stress, we need the highest possible value of W, the work of fracture, in a good stiff material, i.e. one with a high E. It is just because mild steel combines good work of fracture with high stiffness, and is also fairly cheap, that it is so widely used and so important economically and politically.

Although, as we shall see, there are a lot of snags about applying the Griffith equation, which we have just described, and we ought not to regard it as a sort of God-given answer to all design problems, it does in fact do a lot to clarify various structural situations which used to be very obscure and full of mumbo-jumbo.

For instance, instead of messing about with thoroughly bogus 'factors of safety', one can nowadays simply try to design a structure to accommodate a crack of pre-determined length without breaking. The crack length chosen has to be related to the size of the structure and also to the probable service and inspection conditions. Where human life is concerned it is clearly desirable that a 'safe' crack should be long enough to be visible to a bored and rather stupid inspector working in a bad light on a Friday afternoon.

In a really large structure, such as a ship or a bridge, we probably want to be able to put up with a crack at least 1–2 metres long with safety. If we suppose that we want to plan for a crack 1 metre long, then, making the rather conservative assumption that the work of fracture of the steel is 10^5 J/m^2, we find that such a crack will be stable up to a stress of 110 MN/m^2 or 15,000 p.s.i. However, if we want to play safer and plan for a crack 2 metres long, then we shall have to reduce the stress to about 80 MN/m^2 or 11,000 p.s.i.

In fact 11,000 p.s.i. is just about the sort of stress to which large structures are often designed, and in mild steel this stress affords a factor of safety (strictly speaking, what is called a 'stress factor') of between five and six – for what that is worth. As an example of the sort of way this works out in practice, of 4,694 ships subjected to routine inspection in dock, 1,289, or just over a quarter, were found to have serious cracks in the main hull structure – after which, of course, remedial action was taken. The number which actually broke in two at sea, though still too high, was something like one in five hundred, a fairly small proportion. If these ships had been designed to a higher stress, or made of more brittle material, in most cases the cracks would not have been spotted before the ships broke at sea and were lost.

According to the pure and simple Griffith doctrine, a crack shorter than the critical length should not be able to extend at all, and therefore, since all cracks must start life by being short ones, nothing should ever break. In fact, of course, for all sorts of good reasons which are the affair of metallurgists and materials scientists, cracks of less than the critical length do manage to extend themselves, as we shall see in Chapter 15. However, the great point is that they generally do this so slowly that there should be plenty of time to spot them and do something about the situation.

Unfortunately things do not always work out quite that way. Professor J. F. C. Conn, who was Professor of Naval Architecture at Glasgow until recently, told me the story of a cook in a big freighter who was a little startled, when he went into his galley one morning to cook the breakfast, to find a large crack in the middle of the floor.

The cook sent for the Chief Steward, who came and looked at the crack and sent for the Chief Officer. The Chief Officer came and looked at the crack and sent for the Captain. The Captain came and looked at the crack and said 'Oh, that will be all right – and now can I have my breakfast?'

The cook, however, was of a scientific turn of mind, and, when he had disposed of breakfast, he got some paint and marked the end of the crack and painted the date against the mark. Next time the ship went through some bad weather the crack extended a few inches and the cook painted in a new mark and a new date. Being a conscientious man he did this several times.

When the ship eventually broke in two, the half which was salvaged and towed into port happened to be the side on which the cook had painted the dates, and this, Professor Conn told me, constitutes the best and most reliable record we have of the progress of a large crack of sub-critical length.

'Mild' steel and 'high tensile' steel

When a structure fails or seems in danger of failing the natural instinct of the engineer may be to specify the use of a 'stronger' material: in the case of steel, what is known as a 'higher tensile' steel. With large structures this is generally a mistake, for it is clear that most of the strength, even of mild steel, is not really being used. This is because, as we have seen, the failure of a structure may be controlled, not by the strength, but by the brittleness of the material.

Although the measured value of the work of fracture does depend on the way in which the test is done, and it is difficult to get consistent figures, yet the toughness of most metals is undoubtedly reduced very greatly as the tensile strength increases. Figure 13 shows the sort of relationship which exists in simple carbon steels at room temperature.

It is quite easy, and not very expensive, to double the strength of mild steel by increasing the carbon content. If we do so, however, we may reduce the work of fracture by a factor of something like fifteen. In this case the critical crack length will be reduced in the same proportion – i.e. from 1 metre to 6 centimetres – *at the same stress*. However, if we double the working stress, which is presumably the object of the exercise, the critical crack length will be reduced by a factor of $15 \times 2^2 = 60$. So, if a

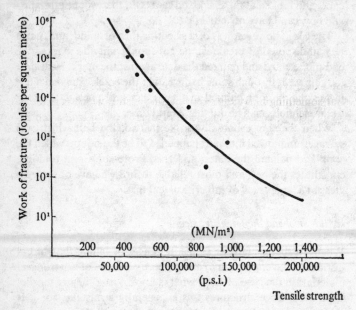

Figure 13.   The approximate relationship between tensile strength and work of fracture for some plain carbon steels. (By courtesy of Professor W. D. Biggs.)

safe crack was originally 1 metre long, it will now measure 1·5 centimetres – which would be thoroughly dangerous in a large structure.

With small components like bolts and crankshafts the situation is different, and it is meaningless to design for a crack a metre long. If we settle for an allowable crack length of, say, 1 centimetre, such a crack may be safe up to a stress of nearly 40,000 p.s.i. (280 MN/m²), and so there is a good case for using a high tensile material. Thus, one consequence of Griffith is that, on the whole, we can use high strength metals and high working stresses more safely in small structures than in large ones. The larger the structure the lower the stress which may have to be accepted in the interests of safety. This is one of the factors which tend to place a limit on the size of large ships and bridges.

The relationship between work of fracture and tensile strength which is sketched in Figure 13 is roughly true for simple commercial carbon steels. It is possible to get rather better combinations of strength and toughness by using 'alloy steels', that is, steels alloyed with elements other than carbon, but these are generally too expensive for large-scale construction. It is for these reasons that something like 98 per cent of all the steel which is made is 'mild steel', that is to say, a soft or ductile metal with a tensile strength of around 60,000 to 70,000 p.s.i. or 450 MN/m².

On the brittleness of bones

> *Children, you are very little,*
> *And your bones are very brittle;*
> *If you would grow great and stately,*
> *You must try to walk sedately.*

R. L. Stevenson, *A Child's Garden of Verses*

But, of course, the bones of children are not very brittle,* and Stevenson was writing rather charming nonsense. In the embryo, bones begin as collagen, or gristle, which is strong and tough but not very stiff (Young's modulus about 600 MN/m²). As the foetus develops, the collagen is reinforced by fine inorganic fibres called osteones. These are formed chiefly from lime and phosphorus and have a chemical formula which approximates to $3Ca_3(PO_4)_2$. $Ca(OH)_2$. In the fully reinforced bone the Young's modulus is increased about thirtyfold to a value of about 20,000 MN/m². However, our bones do not become fully calcified until some considerable time after birth. Naturally, young children are

*There are medical conditions where the bones of quite young people become very brittle, but this state of affairs is rare. An orthopaedic surgeon tells me that the causes are by no means understood.

mechanically vulnerable, but on the whole they tend to bounce rather than break, as one can see on any ski-slope.

However, all bones are relatively brittle compared with soft tissues, and their work of fracture seems to be less than that of wood. This brittleness limits the structural risks which a large animal can accept. As we have already pointed out in connection with ships and machinery, the length of the critical Griffith crack is an absolute, not a relative distance. That is to say, it is just the same for a mouse as it is for an elephant. Furthermore the strength and stiffness of bone are much the same in all animals. This being so, it rather looks as if the largest size of animal which can be regarded as moderately safe is somewhere round about the size of a man or a lion. A mouse or a cat or a reasonably fit man can jump off a table with impunity; it is distinctly doubtful if an elephant could. In fact, elephants have to be very careful; one seldom sees them gambolling or jumping over fences like lambs or dogs. Really large animals, like whales, stick pretty consistently to the sea. Horses seem to present an interesting case. Presumably the original small wild horses did not very often break their bones, but now that man has bred horses big enough to carry him without tiring, the wretched creatures always seem to be breaking their legs.

It is well known that old people are particularly liable to break their bones, and this is generally attributed to a progressive embrittlement of bone with age. No doubt this embrittlement does play some part in causing these fractures, but it does not seem as if it were always the most important factor. As far as I know, there are no reliable data on the change of work of fracture of bone with age, but, since the tensile strength is only reduced by about 22 per cent between the ages of twenty-five and seventy-five, it does not look as if there were a very dramatic reduction. Professor J. P. Paul, of the University of Strathclyde, tells me that his researches seem to indicate that a more important cause of fracture in old people is the progressive loss of nervous control over the tensions in the muscles. A sudden alarm may cause a muscular contraction which is enough to break off the neck of the femur, for instance, without the patient having experienced any external blow. When this happens the patient naturally falls to the ground

—perhaps on top of some obstacle—so that the fracture is blamed, wrongly, on the fall rather than on the muscular spasm. It is said that similar fracture can occur in the hind leg of certain African deer when they are startled by a lion.

Part Two

Tension structures

Chapter 6   Tension structures and pressure vessels

– with some remarks on boilers, bats and Chinese junks

> *That the ship went faster through the water, and held a better wind, was certain; but just before we arrived at the point, the gale increased in force. 'If anything starts, we are lost, sir,' observed the first lieutenant again.*
>
> *'I am perfectly aware of it,' replied the captain, in a calm tone; 'but, as I said before, and you must now be aware, it is our only chance. The consequences of any carelessness or neglect in the fitting and securing of the rigging will be felt now; and this danger, if we escape it, ought to remind us how much we have to answer for, if we neglect our duty.'*

Captain Marryat, *Peter Simple*

The easiest structures to think about are generally those which have to resist only tensile forces – forces which pull rather than push – and, of these, the simplest of all are those which have to resist only a single pull: in other words unidirectional tension, the basic case of a rope or a rod. Although simple unidirectional tension is sometimes to be seen in plants – especially in their roots – the muscles and tendons of animals provide better biological examples and so do vocal cords and spiders' webs.

Muscle is a soft tissue which, when it receives an appropriate nerve signal, is able to shorten itself and so produce tensile forces by pulling in an active way.* However, although muscle is a more efficient device than any artificial engine for converting chemical energy into mechanical work, it is not very strong. So, to produce and sustain any considerable mechanical pull, muscles have to be

---

*The muscular mechanism has recently been understood. It works by feeding energy into edge dislocations which operate, as it were, in reverse. For edge dislocations see *The New Science of Strong Materials*, Chapter 4.

thick and bulky. Partly for this reason muscles are often attached to the bones which they have to manipulate by means of an intervening cord-like tension member made of tendon. Although tendon is unable to contract itself, it is very many times as strong as muscle and therefore needs only a small fraction of the cross-section to take a given pull. Thus the function of tendon is partly that of a rope or wire, although it can also act as a spring, as we saw in the last chapter.

Although some tendons are quite short, many of those in our arms and legs are very long indeed, and they run through the body in almost as complicated a way as the wires of an old-fashioned Victorian bell system. As far as our legs are concerned, muscle is not only bulky but heavy, and the object seems to be to arrange for the centre of gravity of our legs to be as high up in the body as possible. The reason for this is that, in normal walking, the leg operates as a pendulum swinging freely in its own natural period and therefore consuming as little energy as may be. It is because we have to force our legs to oscillate faster than their natural frequency that running is so tiring. The natural period of swing of our legs will be faster the nearer the centre of gravity of the limb is to the thigh-joint. This is why we have thick calves and thighs and, hopefully, small feet and ankles.

However, large feet are not generally so severe a handicap in life's struggle as large hands, whatever people may say about policemen. Our arms, of course, have evolved from front legs, and they seem to have taken the process of remote control even further. Thus, by means of even longer and thinner tendons than exist in our legs, our hands and fingers are operated by muscles located quite a long way away, high up in our arms. So the hand is enabled to have much more slender proportions than would be the case if it had to contain all its own muscles. The advantage of this arrangement mechanically – and perhaps aesthetically – is obvious.

In artificial structures there are a number of simple examples of unidirectional tension, such as fishing lines and loads hanging from cranes. These differ very little from the problem of the brick and the string which we discussed in Chapter 3. However, many of the more interesting cases, such as the rigging of a ship or the design of aerial cableways, are apt to be beset by uncertainties and complications.

In the rigging of a ship there would, of course, be no difficulty about determining a safe thickness for each rope, provided only that one knew what loads they would have to carry. The difficulty lies in predicting the magnitudes of the many different forces which operate in so complicated an affair as a sailing ship. Although there are several ways in which one might set about this, I strongly suspect that most yacht designers prefer to rely on what might be described as experienced guesswork. However, it is just as well to get one's guesses right, since the failure of a vital piece of rigging is likely to result in the loss of a mast. If this happens when the ship is caught on a dangerous lee shore, like Marryat's frigate, then the consequences will be serious.

Nowadays ski-ing is a vast international industry which is dependent upon the reliability of many thousands of cable-cars and ski-lifts. I suppose that most of us have worried, in our more vertiginous moments, about the strength of the wire ropes which support chair-lifts and cable-cars above what seem to be rather frightening chasms. Actually, accidents very seldom occur directly from the failure of one of these cables in tension. This is because in this case the static loads are known pretty accurately, and it is not difficult to do the sums and ensure an ample factor of safety. More serious risks arise from such matters as the excessive swaying of the cables in the wind, so that the cars are likely to strike each other as they pass or perhaps to hit the supporting pylons. Here again, designers seem to rely mainly on precedent and guesswork.

A very different application of unidirectional tension theory is concerned with the strings of musical instruments. The frequency*

*The number of vibrations per second (i.e. the frequency), n, of a stretched string can be written:

$$n = \frac{1}{2l} \sqrt{\frac{s}{\rho}}$$

where l = length of string (m)

ρ = density of material from which string is made (Kg/m³)

s = tensile stress in string (N/m²)

of the note given out by a stretched string depends, not only on its length, but also upon the tensile stress in it. In stringed instruments the appropriate stresses are produced by stretching the strings – which are made of stiff material, such as steel wire or catgut – across a suitable framework, which may be the wooden body of a violin or the cast-iron frame of a piano. Since both the strings and the framework are stiff, very small extensions greatly affect the stress in a string and therefore the frequency of its note. This is why such instruments are so sensitive to 'tuning'. It is also why one can use the note emitted by a rope when it is 'twanged' as an indication of the stress in the material. The Roman army used to require that the officers in charge of military catapults should have a good musical ear, so that they could assess the tensions in the tendon ropes of these weapons when they were set up and tuned for action.

Although the human voice differs in many ways from a stringed instrument, somewhat similar considerations apply to it. The mechanisms of voice production are rather complicated, but our larynx plays an important part in both singing and talking. It may be interesting to note that the various tissues of the larynx are among the few soft tissues in the body which conform approximately to Hooke's law; most of the other body tissues obey quite different and rather weird laws of their own when they are stretched, as we shall see in Chapter 8.

The larynx contains the 'vocal cords', which are strips or folds of tissue whose tensile stress can be varied by muscular tension so as to control the frequency with which they vibrate. Because the Young's modulus of the vocal folds is rather low, large strains sometimes have to be applied to them in order to cause the necessary stresses; they are, in fact, stretched by a good 50 per cent when we want to achieve the top notes.

Incidentally, the higher frequencies of the voices of women and children are caused, not by higher tensions in their vocal cords, but simply by the fact that the larynx is smaller and the vocal cords therefore shorter. There is a surprising difference in this respect between grown-up men and women, the relevant larynx measurements being about 36 millimetres for men against about

26 for women. However, the larynxes of both boys and girls are of very similar size up to the age of puberty. The 'breaking' of boys' voices is due, not to any change of tension in the cords, but to a rather sudden enlargement of the larynx around the age of fourteen.

Pipes and pressure vessels

Plants and animals might be regarded to a considerable extent as so many systems of tubes and bladders whose function is to contain and to distribute various liquids and gases. Although the pressures in biological systems are not usually very high, they are by no means negligible, and living vessels and membranes do burst from time to time, often with fatal results.

In technology the provision of reliable pressure vessels is a fairly modern achievement and we seldom stop to think how we should get on without using pipes. For the lack of pipes capable of conveying liquids under pressure the Romans incurred enormous expenses in building masonry aqueducts upon tall arches in order to carry water in open channels across miles of undulating country. The earliest approximations to pressure-tight containers were the barrels of guns, and, historically, these were never very satisfactory and quite frequently failed. A list of the people who have been killed by the accidental bursting of guns, from King James II of Scotland downwards, would be long and impressive. Nevertheless, when gas lighting began to be installed in London, soon after 1800, the pipes had to be made by Birmingham gunsmiths, and in fact the earliest gas-pipes were actually made by welding musket-barrels end to end.

Although there are innumerable accounts of the history of the steam-engine, relatively little has been written about the development of the pipes and boilers on which it depended and which, in reality, presented more difficult problems than the actual mechanism. The earliest engines were very heavy and bulky and consumed vast amounts of fuel, chiefly because they worked at very low steam pressures – which was perhaps just as well in view of the nature of contemporary boilers.

The production of engines which were light, compact and altogether more economical was wholly dependent on the use of much higher working pressures. In the steamships of the 1820s, with steam pressures of about 10 p.s.i. – provided by square 'haystack' boilers – the coal consumption was around 15 lb. weight per horsepower hour. In the 1850s engineers were still talking in terms of 20 p.s.i. and about 9 lb. per horse-power hour. By 1900, pressures had gone up to well over 200 p.s.i., and coal consumption had fallen to 1·5 lb. per horse-power hour – a tenfold reduction in eighty years. It was not the steamship, as such, which drove sailing ships from the high seas, but the *high-pressure* steamship with triple-expansion engines, 'Scotch' boilers, low fuel costs and long range.

The high-pressure boiler was not developed without incident. Throughout most of the nineteenth century boiler explosions were relatively frequent and the consequences were sometimes very terrible. The American river steamers, in particular, were pioneers of high-pressure working. During the middle years of the century the Mississippi steamboats used regularly to indulge in dramatic races over thousands of miles of river. The designers of these vessels were prepared to sacrifice almost everything to speed and lightness, and they took what might charitably be called an optimistic approach to boiler design. As a result, during the years 1859–60 alone, twenty-seven of these ships were lost as a result of boiler explosions.*

Although some of these accidents were due to criminal practices such as the tying down of safety valves, most of them were basically caused by lack of proper calculations. This was a pity because in fact the basic calculations needed to determine the stresses in simple pressure vessels are very easy – so easy indeed that, as far as I can find out, nobody has ever bothered to claim the credit for originating them, and only the most elementary kind of algebra is required. †

* But then, during the same period, eighty-three steamboats were destroyed by fire, eighty-eight by running into sunken trees, and seventy from 'other causes'. It seems that Life on the Mississippi, in the showboat days, was not uneventful.

† A partial solution was provided by Mariotte around 1680, but of course he was unable to make use of the concept of stress.

Spherical pressure vessels

As soon as we come to consider any kind of pressure vessel or container – which includes such things as balloons and bladders and stomachs and pipes and boilers and arteries – we have to deal with tensile stresses which operate in more than one direction at the same time. This may possibly sound complicated but presents, in fact, no cause for alarm. The skin of any pressure vessel really performs two functions. It has to contain the fluid by being water-tight or gastight, and it has also to carry the stresses set up by the internal pressure. Nearly always this skin or shell is subjected to tension stresses acting in both directions in its own plane, that is to say, parallel to its surface. The stress in the third direction, per-pendicular to its surface, is usually negligibly low and can be forgotten about.

It is convenient to look first at pressure vessels of spherical shape. The skin or shell of the bladder-like object in Figure 1 is supposed to be reasonably thin, say less than about a tenth of the diameter. The radius of the shell, taken at the middle of the wall thickness, is r. The thickness of the wall or shell is t and the whole thing is subject to an internal fluid pressure of p (all these being in whatever units we happen to patronize).

If we imagine that we slice the thing in two, like a grapefruit, then from Figures 1, 2 and 3 it is pretty clear that the stress in the shell – in all directions parallel to its own surface – will be

$$s = \frac{rp}{2t}$$

This is a useful practical result, and it is in fact a standard engineering formula.

Cylindrical pressure vessels

Spherical containers have their uses, but clearly cylindrical vessels have wider applications, especially to things like pipes and tubes. The surface of a cylinder has no longer the same sort of symmetry as that of a sphere and so we cannot assume that the stress along a cylinder is the same as that around its circumference; and in fact

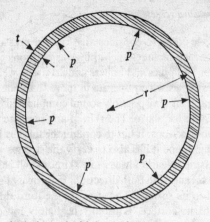

Figure 1. A spherical vessel with internal pressure *p*, mean radius *r*, and wall thickness *t*.

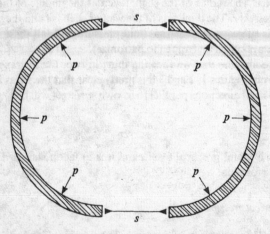

Figure 2. Imagine the vessel sliced in two across any diameter. The resultant of all the pressure forces acting on the inside of each half of the shell must equal the sum of all the stresses which would have acted on the cut surface, whose area is $2\pi r t$.

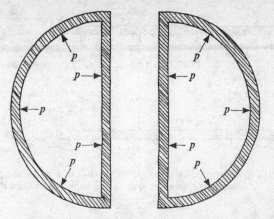

Figure 3. The resultant of all the pressure forces acting on the inside curved surface of a hemisphere will be equal to the same pressure acting on a flat disk of the same diameter, which must be $\pi r^2 p$. Hence

$$\text{the stress } s = \frac{\text{load}}{\text{area}} = \frac{\pi r^2 p}{2\pi r t} = \frac{rp}{2t}$$

it isn't. Let us call the stress in the shell along the length of the cylinder s_1 and that around the circumference of the shell s_2.

From Figure 4 we can see that the stress along the shell, s_1, must be the same as that in the sphere which we have just been considering, that is to say

$$s_1 = \frac{rp}{2t}$$

To get at s_2, the circumferential stress in the shell of the cylinder, we now slice, in our imaginations, in the other plane, after the fashion of Figure 5; from which we see that

$$s_2 = \frac{rp}{t}$$

Thus the circumferential stress in the wall of a cylindrical pressure vessel is twice the longitudinal stress, i.e. $s_2 = 2s_1$ (Figure 6). One consequence of this must have been observed by everyone who has ever fried a sausage. When the filling inside the sausage swells

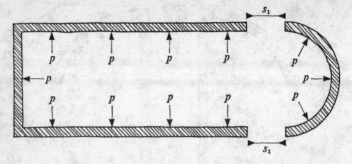

Figure 4. The longitudinal stress, s_1, in the shell of a cylindrical pressure vessel is the same as that in the equivalent spherical vessel.

$$s_1 = \frac{rp}{2t}$$

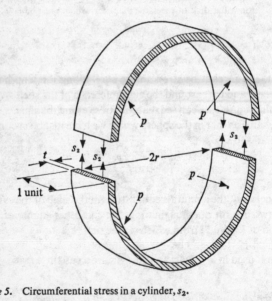

Figure 5. Circumferential stress in a cylinder, s_2.

$$s_2 = \frac{rp}{t}$$

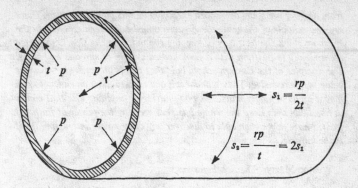

Figure 6. Stress in the wall of a cylindrical pressure vessel.

and the skin bursts, the slit is almost always longitudinal. In other words, the skin has broken as a consequence of the circumferential, not the longitudinal, stress.

These sums are continually cropping up in engineering and in biology. They are used to calculate the strength of pipes and boilers and balloons and air-supported roofs and rockets and space-ships. As we shall see in Chapter 8, the same simple piece of theory applies to the whole question of development from amoeba-like forms of life towards more elongated and mobile primitive creatures.

Another consequence of the algebra we have just done is that to contain a given volume of fluid at a given pressure will require a greater weight of material if we use a cylindrical vessel than if we use a spherical one. Where weight is very important – as it is with the oxygen-bottles which climbers use at high altitudes and also with aircraft starter bottles – then spherical vessels are usual. For most other purposes, where weight is not so serious a matter, cylindrical bottles are cheaper and more convenient. The 'gas-cylinders' used in hospitals and garages are a case in point.

Chinese engineering – or better bulge than bust

There is an interesting problem which has to be solved by the designer of every sailing vessel. It is: What is the best method of preventing the ship from flinging her spars overboard? Opinion on this point is divided. There are two schools of thought: the Eastern and the Western schools. In the West we think the best way of keeping the masts in the ship is to fasten them rigidly in position with a complicated system of shrouds and stays. The disciples of the Eastern school hold that this is all nonsense – besides being very expensive. They stand a tall and rickety mast on end, set upon it vast areas of gunny mats, bamboo matting or anything else that comes to hand and then keep the whole business erect by the power of faith. At least, I have never been able to discover any other power interesting itself in the miracle.

Weston Martyr, *The Southseaman*

The theory of pressure vessels, which we have just derived, also applies, with minor modifications, to things other than closed containers: that is, to 'open' membranes and fabrics which have to sustain pressure from the free movement of wind or water. Of such a nature are tents and kites and awnings and fabric-covered aircraft and parachutes and the sails of ships and windmills and eardrums and fishes' fins and the wings of bats and pterodactyls and the sails of the jellyfish called Portuguese men-of-war.

For all such purposes it is expedient and economical (as we shall see in Chapter 14) not to use a 'rigid' panel or shell or monocoque but to cover a stiff open framework of rods or spars or bones with some kind of flexible fabric or skin or membrane. Such a structure cannot be quite rigid, and it will be realized that, as soon as any lateral force comes upon the membrane by reason of the pressure of wind or water, it must deflect or bow into a curved shape which, to a first approximation, may be treated as a part or segment of a sphere or cylinder, and so the stresses in the membrane will obey much the same laws as those in the shell of a pressure vessel.

From this it is very easy to show that the force or tension in the membrane, per unit width, is pr, the product of the wind pressure (p) and the radius of curvature (r) of the membrane. Thus the more sharply the membrane is curved the less the force in it will be, and so the load which it puts upon the supporting framework will also be diminished.

When the wind blows, the pressure caused by the wind increases as the square of the wind velocity. In a strong wind the pressure becomes very high indeed and so do the loads upon the supporting structure. According to our Western, engineering-school, way of thinking there is very little we can do about this, for we would rather be seen dead than allow the membrane – whether it be a sail or part of an aeroplane or whatever – to bulge appreciably between its supports. Of course, we can never manage to keep the fabric perfectly flat, but we do everything we can to keep it as taut as possible. What we actually do is to make the supporting framework strong and heavy and expensive and hope that it won't break – which of course it often does.

For instance, the rig which has been developed for modern racing yachts generally consists of tubular metal spars and almost inextensible Terylene sails. This aerodynamic mechanism is kept up to its job by many ropes and wires which, in turn, are tautened to a frightening degree by means of screws and winches and hydraulic jacks, all of which are needed to cope with the enormous loads set up in the sails when the vessel is sailing fast in a breeze of wind. The whole thing is a miracle of engineering 'efficiency' but it is also horribly expensive. Ships of this sort convey to their occupants a feeling of tenseness which is anything but restful.

A simpler and cheaper way of doing the job is to arrange for the sail to bulge between its supports so that, as the wind pressure

Figure 7.   Chinese junk rig.

increases, the radius of curvature diminishes, and so the tension force in the canvas remains roughly constant however hard the winds may blow. Naturally, one has to ensure that the distortions which help to ease the structural problem do not create aerodynamic ones.

One elegant and satisfactory way of doing this has been devised by the Chinese, who, after all, have been sailing about the seas in moderate comfort and safety for a good many centuries. The rig of the traditional Chinese junk varies according to local custom but is generally very much like Figure 7. The battens which cross the sails are attached to the masts and, since the whole rig is constructed from flexible materials, as the wind increases, the sail bows out between the battens after the fashion of Figure 8 without

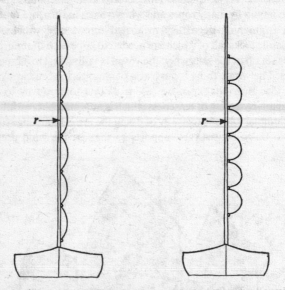

Figure 8.   Edge-on view of a junk sail with halyard eased.

much loss of aerodynamic efficiency. If it doesn't bulge enough it is quite simple to ease the halyard until it does. Lately, Colonel 'Blondie' Hasler (of Bordeaux Raid fame) has taken up the Chinese lug sail with very satisfactory results. Several yachts with

Colonel Hasler's rig have made long ocean voyages with success and in a comparatively relaxed manner. The 'hang-gliders' which are now so popular are designed on much the same principles, and, although they may shock the traditionalists, they are cheap and strong and they do seem to work.

Bats and pterodactyls

> *Take from the goblin his crinkly face,*
> *His pointed ears from the gnome;*
> *Borrow the nose of a leprechaun*
> *And smuggle it carefully home;*
> *Sew bawkie fingers to banshee wrist;*
> *Stitch gossamer vellum between;*
> *Fit legs to straddle with knees atwist*
> *From a body of velveteen.*

Douglas English (*Punch*, 11 July 1923)

The resemblance between a bat and a Chinese junk is immediately obvious (Figure 9). In all bats the wings are constructed by stretching a membrane of very flexible skin over a framework of

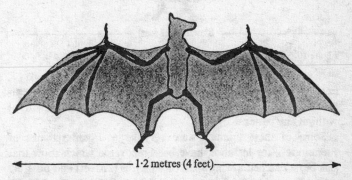

←———————— 1·2 metres (4 feet) ————————→

Figure 9. The fruit-bat.

long, thin bones which are, in essence, the fingers of a hand. Fruit-bats, for instance, are quite large animals with a wing-span of four feet or something over a metre. In their native India, where they are a pest, they think nothing of flying thirty or forty miles in a night in order to rob an orchard. Since they can do this without becoming unduly exhausted they must therefore be efficient flying machines. Furthermore, to save weight, as well as what is called 'metabolic cost', they have gone a long way in the matter of cutting down the thickness of their wing-bones.

When a fruit-bat is photographed in flight it can be seen that, on the down-stroke of the wing, the skin membrane bulges upwards into a form which is roughly semi-circular, thereby minimizing the mechanical load upon its bones. It is clear that there can in practice be little or no aerodynamic loss as a consequence of this change of shape.

About 30,000,000 years ago the place of birds was largely filled by a wide range of flying creatures called Pterodactyls (finger-wings). Many of these much resembled bats, except that only one

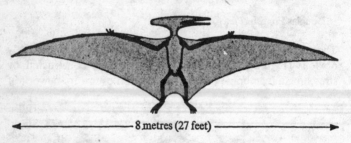

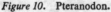
← ———————— 8 metres (27 feet) ————————→

Figure 10.   Pteranodon.

finger, the little finger, played any structural part. So the membrane wing of pterodactyls was rather like a Bermuda mainsail without any battens.

Some of these animals were very large. Fossil remains of Pteranodon, for instance, have been recovered which show that this beast reached a wing-span of 8 metres (27 feet) and possibly more. It stood about 3 metres high (10 feet), and it seems that its total weight was probably only about 20 kilograms (44 lb.). There was therefore little weight available either for bony structure or for flying muscles. Recently, the discovery of even larger pterodactyls, about twice the span of Pteranodon, has been reported from America.

Pteranodon was probably pelagic: that is to say, it filled, roughly speaking, the ecological niche which is now occupied by the albatross. Like the albatross, it seems to have lived mostly in the air, soaring close above the deep-sea waves and fishing on the wing. Even more than the fruit-bat, the wing bones of Pteranodon appear, from the fossils, to have been almost unbelievably thin and fragile. Of course, we know nothing experimentally about the elasticity of the skin which covered these vast wings, but it seems fair to infer that this skin must have behaved very much like that of a bat. The aerodynamic efficiency of the whole system must have been high and comparable to that of the modern albatross.

Why do birds have feathers?

Although bats flourish and survive very well today, pterodactyls were superseded by birds, which have feathers, a great many years ago. It is possible, of course, that the extinction of pterodactyls had nothing to do with structural considerations, but it is also possible that there is something special about feathers which gives birds an edge over other flying creatures. When I worked at the Royal Aircraft Establishment I used to ask my superiors, from time to time, whether it would not perhaps be better if aeroplanes had feathers; but I seldom succeeded in extracting a rational or even a patient answer to this question.

But, after all, why *do* birds have feathers? Given the job of designing a flying animal, a modern engineer would perhaps produce something like a bat, or possibly some sort of flying insect. I do not think that it would occur to him to invent feathers. Yet presumably there are very good reasons for their existence. One imagines that both bats and pterodactyls tend to lose a good deal of energy in the form of heat from the skin of their wings; but then reasonable heat insulation could be provided by fur.

Perhaps this is what did happen at an early stage in the evolution of birds, because feathers, like horns and claws, developed from hair. However, hair is presumably better when it is soft, and so the keratin from which hair is made has quite a low Young's modulus. In feathers the keratin molecule has been made stiffer by cross-linking the molecular chains with sulphur atoms (which accounts for the smell of burnt feathers).

There are, no doubt, aerodynamic advantages in using feathers, since their employment extends the choice of outside shapes which the animal can make use of. For one thing, 'thick' wing-sections have often better aerodynamic efficiencies than the thin ones which result from membranes. It is easy to get an efficient 'thick' section by padding out the wing profile with feathers at the cost of very little weight increase. Furthermore, feathers are better adapted than skin and bone for providing anti-stalling devices such as 'slots' and 'flaps'.

However, I am inclined to think that the main advantage of feathers to an animal may be structural. Anybody who has flown model aeroplanes knows, to their cost, how vulnerable any small flying machine must be to accidental damage from things like trees and bushes, or even from careless handling. Many birds fly constantly in and out of trees and hedges and other obstacles. Indeed they use such cover as a refuge from their enemies. For most birds the loss of a reasonable number of feathers is not a very serious matter. Besides, it is better to leave the cat with a mouthful of feathers than to be eaten.

Feathers not only enable birds to get away with more local scrapes and abrasions than other animals, but the body of the bird is protected from more serious damage by its thick resilient armour. The Japanese feather armour which one sees in museums was not, as one might suppose, the picturesque nonsense of a primitive people who did not know any better. It was an effective protection against weapons like swords. In the same kind of way, during the Russo-Finnish war, Finnish armoured trains were protected by bales of paper; and modern fighter-pilots' splinter-proof boots are made from many layers of Cellophane. When a hawk kills a bird in the air it does not usually do so by wounding it with its beak or talons – which would probably not penetrate the feathers. It kills by striking the bird in the back with its out-stretched feet so as to impart a violent acceleration to the bird as a whole which has the effect of breaking its neck – very much as happens in judicial hanging.

The whole constitution and design of feathers seem to be

extremely cunning. Feathers probably do not need to be especially strong, but they do need to be stiff and at the same time resilient and to have a high work of fracture. The work of fracture mechanism of feathers is something of a mystery; at the time of writing I do not think anybody knows how it works. Like so many work of fracture mechanisms, that of feathers is sensitive to what appear to be small changes. Everybody who has kept and flown hawks knows that these intelligent, exacting and maddening birds lose condition very easily. Even when they are properly fed and exercised in captivity, hawks' feathers are very apt to become brittle and break off with undue frequency. The cure or palliative for this is to join the broken parts of the feather together again by 'imping'. This is done by inserting a double-ended 'imping-needle', with a little glue, inside the hollow of the shaft at the break. The details of this process are described in the sixteenth-century books on hawking.

When one considers the appalling and expensive frequency with which motor cars nowadays incur bumps, bashes and abrasions, one sometimes wonders whether they have not a lesson to learn from the birds. Incidentally, I am told that, since the American army practically lives on chicken, there exist somewhere in the United States enormous quantities of unwanted chicken feathers. It would be rather nice to find a use for them.

Chapter 7   Joints, fastenings and people

– also about creep and chariot wheels

> *And here I want to tell you a story about a ship that was made during the war. She was a steamer, and she was built of wood – good wood; and the men who designed her were good and able craftsmen too . . .*
>
> *She went along like a man who carries too heavy a burden, and presently she tripped and stumbled (it was only a little ground-swell) – and she opened out and fell apart like a flimsy old crate that someone had stepped on. In five minutes there was nothing there at all except a floating scum of coal dust, with some timbers and an odd man or two bobbing about in the middle of it.*
>
> *This is a true story; but the point I want you to notice is that this ship was made by carpenters: house carpenters – shore carpenters; and she was not built by shipwrights at all.*

Weston Martyr, *The Southseaman*

The steamer in Weston Martyr's story sank, rather suddenly, because the joints which were supposed to hold her timbers together were much too weak, although the house carpenters who built her – who were honest men in their own way – were presumably satisfied with them. In fact, when a shore carpenter is building a house or putting together traditional furniture he is in the habit of making joints which a naval architect or an engineer would regard as weak and highly inefficient. Weak these joints certainly are; whether they are 'inefficient' depends upon what one is trying to do. The purposes of a builder of houses may not be at all the same as those of a builder of ships or aeroplanes.

It is perhaps too often assumed by engineers that an 'efficient' structure is always one in which each component and each joint is exactly strong enough for the loads which it has to bear, so that, for a given strength, the smallest amount of material is used and the weight is minimized. Such a structure would, ideally, be equally likely to break anywhere. Or indeed, like the 'one-hoss shay', it might break everywhere at once. To work towards this kind of efficiency calls for great vigilance on the part of the engineer, since the least fault in design or manufacture must introduce a dangerous weakness.

Approximations to this kind of structure do, of course, exist, especially in ships and aeroplanes and in some kinds of machinery where weight-saving is very important. However, this represents an unduly specialized way of looking at the problem of efficiency, and it takes no account of the need for rigidity, let alone of the need for economy. Structures of the one-hoss shay type are sometimes necessary, but they are always expensive both to build and to maintain. Weight-saving by means of structural perfectionism is one of the factors which make space travel such an extravagant luxury. Even at a mundane level we may reflect that the cost of usable space, per cubic metre, is about twenty times as high in a small ship as it is in an ordinary house; the cost of space in aircraft is a great deal higher still.

Builders and joiners have more sense than to go in for fancy structures of this kind; houses are quite expensive enough as it is, and these people know very well that in the great majority of the common or stationary affairs of life the design of a structure is influenced much more by its stiffness than by its strength.

Indeed it is the relative importance of the need for strength and for stiffness which really lies at the root of the question of the cost and efficiency of structures. Where the need is chiefly for rigidity rather than strength, the whole problem becomes very much easier and cheaper. This is nearly always the case with furniture and floors and staircases and buildings generally, and also with cookers and refrigerators and with many tools and heavy machinery and with some of the parts of motor cars. These things do not very often break, but, if we made the material much thinner, the deflections and bendiness and general wobbliness would soon become unacceptable. Thus, to be rigid enough, the various parts generally have to be so thick that the stresses in them are very low, often, from the engineer's point of view, absurdly so.

It follows that, in structures of this kind, even if the material is riddled with defects and stress concentrations, it probably does not matter very much, and, what is more, the strength of the joints is unlikely to be critical; in many cases, a few nails may be perfectly adequate. This sort of thing is, of course, the basis of most people's instinctive approach to design. Millions of people who have never heard of Hooke's law or Young's modulus can guess the stiffness of a table or a chicken-coop quite nearly enough by experience and common-sense, and, if such things are made stiff enough, they are very unlikely to break under their ordinary, everyday loads.

Furthermore, a little bit of 'give' in some of the joints may be no disadvantage, and this is more likely to be available in a traditional joint than in a sophisticated one. For one thing a certain amount of flexibility may enable the loads to be evened out in a beneficial way. Although it is true that furniture does not very often get broken, quite a good way of attempting to do so is to sit on a chair, three of whose legs are on the carpet while the fourth rests, hopefully, on the bare floor. With traditional furniture the load may be spread over all the four legs by the distortion of the tenoned joints; in modern factory-made chairs with 'efficient' glued joints, these joints may just break, after which the chair is difficult to repair in any satisfactory way.

Another reason for encouraging a certain amount of flexibility in joints is that wood, and sometimes other materials, change their dimensions with the weather. Wood shrinks and swells in the cross-grain direction by up to 5 or even 10 per cent. Traditional joinery allows for this by means of 'inefficient' slotted joints. In Churchill College we had a fine new High Table made from the best and most expensive wood, which had been scientifically glued together with strong, rigid joints. After a few months in the scientifically heated Hall, this table shrank and split down the middle. The result was not an unobtrusive little crack but a crevasse many yards long and quite capable of providing sheltered accommodation for large numbers of peas of normal or standard diameter.

Strong joints and frail people

Many deflection-controlled peasant structures are wholly excellent in their proper places but when we come to demand weight-saving and strength and mobility we may get into all sorts of difficulties, especially in relation to the reliability of the joints between the various parts. Historically, this has always been the most serious problem in ship construction and in windmills and water-mills. The great skill of the old shipwrights and millwrights lay in somehow combining sufficient strength for safety with the modicum of flexibility needed to allow for the 'working' of timber. The older shipwrights erred on the side of flexibility, and, though their ships were often excessively leaky, they seldom actually broke at sea. It required the administrative abilities of modern war-time governments to produce wooden ships which really did fall to pieces.

Troubles with joints in ships and aircraft were a fairly prominent feature of both the World Wars. During the first war the Americans built a large number of wooden ships, both steam and sail, frequently by unorthodox methods; and many of these ships broke up. In the second war they produced even greater numbers of welded steel steamers, of which an even higher proportion broke, either at sea or in harbour. In England, in both wars, we manufactured very large quantities of wooden aircraft, which always seemed to be having joint troubles of one kind or another.

As far as aircraft are concerned this was not wholly surprising, for I remember being shown, right inside vital glued joints in the main structure, on various separate occasions:

1. A pair of scissors.
2. A first-aid manual (pocket-size).
3. No glue at all.

On the whole I do not think that most of these accidents were caused by sub-normal or abnormal people; I am afraid the guilt generally lay with very ordinary people, and that was just the trouble. Naturally, people get tired or bored, but I think the root of the matter was much deeper than that. Very few of those who made, or failed to make, these joints had any personal experience of a situation in which the failure of a joint could cause a fatal accident, though collectively they had a great deal of experience of things like cupboards and garden sheds, where the strength of the joints really mattered very little. All our efforts to persuade them that a badly made joint was morally equivalent to manslaughter foundered on a deeply-held folk tradition that it was silly to fuss about such things and that strength is a boring subject anyway. All this would not have mattered so much if it had not been practically impossible to inspect the joints properly after they were made.

In more recent years very efficient metal-to-metal adhesives have been developed which have a number of solid technical advantages, always provided that the joints are really conscientiously made. Unfortunately, their use in modern aircraft has been handicapped by the fact that it has proved necessary to provide a separate inspector to watch each worker throughout the gluing operation – also inspectors to inspect the inspectors. Rather naturally, these arrangements have proved expensive. In spite of all this, I am told that the use of glue in modern metal aircraft is increasing.

Stress distribution in joints

Since the function of a joint is to transmit load from one element of a structure to its neighbour, stress has somehow got to get itself out of one piece of material and then get itself into the adjoining

Figure 1.   Glued scarf joint in timber.

piece; such a process is only too likely to result in severe concentrations of stress and consequent weakness. All the same, in a few favourable circumstances it may be possible to arrange for the stress to pass uniformly across the joint from one component to the other with little or no concentration of stress; this is more or less the case with a glued scarf joint in timber (Figure 1) and a butt-weld in metal (Figure 2).

Figure 2. Butt-weld in metal.

However, it is by no means always practicable to use scarfed or butt-welded joints, and some form of lapped joint between two adjacent planks or plates is probably more common. This sort of geometry at once introduces stress concentrations, and as far as a 'rigid' lapped joint is concerned it does not make much difference whether the joint is glued, nailed, screwed, welded, bolted or riveted. In all cases most of the load is transferred at the two ends of the joint (Figure 3).

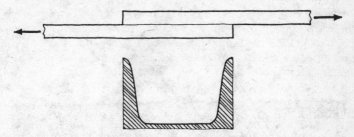

Figure 3. Load transfer in a lapped joint.

For this reason the strength of such joints depends largely upon their width and very little upon the length of the overlap between the parts. This is why the simplest and commonest forms of riveted

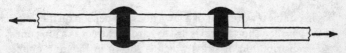

Figure 4. Riveted lapped joint.

and welded joints between two metal plates (Figures 4 and 5) are reasonably effective and not much improved by complicating them.

Very often we want to provide an end attachment for a tension bar or rod to some kind of socket or solid anchorage; again much the same considerations apply, except that in this case there is only one stress concentration, which usually occurs at the point where

Figure 5.   Welded lapped joint.

the rod enters its socket (Figure 6). If the rod is screwed into its anchorage, for instance, nearly all of the load is taken out by the first two or three threads, and any extra length of rod within the socket will do little or no good. Thus the difficulty which a thrush

Figure 6.

has in pulling a worm out of a lawn does not depend on the length of the worm; a short worm is just as hard to extract as a long one.*

The distribution of stress which is shown in Figure 6 applies when the two components of the joint have similar Young's

---

*Note that, if an 'undrawn' nylon thread be cast into a block of 'rigid' plastic, the thread can always be drawn out of the plastic by pulling on it, *however long the thread may be*. This is a good way of making long and complicated holes, for instance in wind-tunnel models, for pressure measurements.

moduli, which is usually the case with metal-to-metal joints. It also applies when the rod or tension bar is less stiff than the material of its socket or anchorage – which appears to be the case with worms and lawns. If the rod or bar is substantially stiffer than the material into which one is trying to anchor it, however, the stress situation may be reversed and the stress concentration may exist mainly at the bottom or inner end of the rod or insert (Figure 7).

In practice, of course, both situations are likely to weaken the joint about equally. There may exist, perhaps, a ratio between the modulus of the insert and that of its surroundings which would give an optimum distribution of stress in the joint; but, if there is such a ratio, it is very difficult to hit it off in real life.

At one time I was concerned with making point attachments

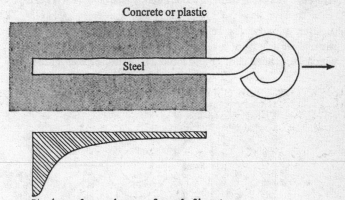

Concrete or plastic

Steel

Load transfer maximum at far end of insert

Figure 7. Load transfer in embedded rods under tension.

between a reinforced plastic wing and the metal fuselage of an aircraft. Although I should have known perfectly well about stress concentrations and worms in lawns and so on, I was foolish enough to begin by moulding strong wire cables, with frayed-out ends – like the roots of a tree – into the body of the plastic. When specimens of this ill-conceived construction were loaded in a testing machine, the wires pulled out of the plastic with a succession of cracking noises and at ridiculously low loads.

In the next experiment sword-like tapered steel blades or prongs were substituted for the cables and were moulded into the plastic wing structure after being coated with a suitable adhesive (Figure 8). This time the test-specimen failed, not with a series of cracking noises, but with one loud bang, but still at just as low a load.

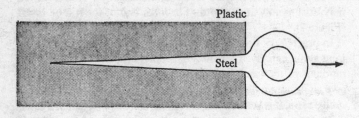

Figure 8.   The wrong shape for a steel insert. This arrangement is weak.

After a pause for reflection and intelligent thought about worms, we tried out a series of wide spade-shaped steel inserts which were much shorter and looked something like Figure 9. All

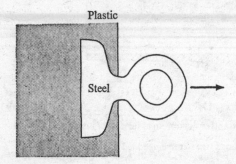

Figure 9.   The right shape for a steel insert. This is much stronger.

these failed at far higher loads which were, in each case, proportional to the breadth of the 'spade'. By developing this design we were able to take out loads in the region of 40 to 50 tons from plastic structures by means of quite a small steel fitting.

Such joints depend entirely upon the adhesion between the metal and the plastic and must therefore be moulded conscientiously and under suitable inspection. They must also be designed with care, because, in all such cases, adhesion between a metal and a non-metal will fail completely as soon as the metal reaches its yield-point and ceases to behave elastically.* Since the stresses in the metal are much higher than one might expect, it is generally necessary to make the insert from high tensile steel, carefully heat-treated. Furthermore the 'trailing edge' of the steel insert must be ground sharp, like a chisel.

Riveted joints

'I've got one-fraction of an inch of play, at any rate,' said the garboard-strake, triumphantly. So he had, and all the bottom of the ship felt easier for it.

'Then we're no good,' sobbed the bottom rivets. 'We were ordered – we were ordered – never to give; and we've given, and the sea will come in, and we'll all go to the bottom together! First we're blamed for everything unpleasant, and now we haven't the consolation of having done our work.'

'Don't say I told you,' whispered the Steam, consolingly; 'but, between you and me and the last cloud I came from, it was bound to happen sooner or later. You had to give a fraction, and you've given without knowing it. Now hold on, as before.'

Rudyard Kipling, *The Ship that Found Herself*

Riveted joints in steel structures are rather out of fashion, chiefly because they are expensive but partly because they tend to be heavier than welded joints. This is a pity, because riveted joints have several advantages. A riveted joint is reliable and easy to inspect, and in a large structure it acts to some extent as a crack-stopper: that is to say, if a really large and healthy Griffith crack gets under way, it may quite often, though not infallibly, be stopped or delayed by the moat or discontinuity of a riveted joint.

Even more importantly, riveted joints can slip a little and so redistribute the load, thus evading the consequences of the stress

---

*This is also true for the adhesion between metals and paint or enamel, including 'vitreous enamel', i.e. glass. Before the days of modern extenso-meters, engineers used to judge the 'yield-point' of hot-rolled steel by the load at which the 'mill-scale', or black oxide film, cracked off the surface.

concentrations which are the bane of all joints. The process has been described for all time in *The Ship that Found Herself*, and indeed Kipling's feeling for the problems of stress concentrations and cracks in structures, many years before Inglis and Griffith, is very remarkable; some of his stories about structures might well be required reading for engineering students.

Because each individual rivet can slip very slightly, the worst effects of stress concentrations may be reduced, and so it may be worth while to make lap joints having several rivets in series,

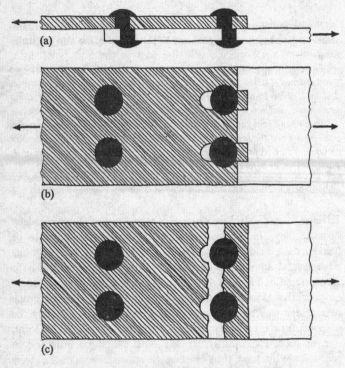

Figure 10. Three of the ways in which a riveted joint may fail.
  (a) Failure by shearing the rivets.
  (b) Failure by tearing the rivets out of the plate (i.e. by 'bearing' or elongation of the holes).
  (c) Failure by tearing the plates.

since the end rivets may be able to slip enough to enable those in the middle to do some work. When a newly made riveted joint between steel or iron plates has settled itself into a reasonable distribution of load, then rust may have a chance to play its beneficent part. The products of corrosion, iron oxides and hydroxides, expand and so lock the joint and prevent it from sliding backwards and forwards when the load is reversed. Furthermore, the rust transmits some of the shearing forces between the plates, rather like a glue, and therefore the strength of a riveted lap joint generally increases with age.

When rivet holes are made in large steel structures, such as ships and boilers, it is usual to punch them. Although this is a quick and cheap way of making holes in steel it is not entirely satisfactory, since the metal at the edge of the hole is left in a brittle condition and also often contains small cracks. Since there will certainly be stress concentrations in this region, this is not a good state of affairs. For this reason, in high-class work, it is usual to punch the holes under-size and then ream them. Although this adds to the expense, it also adds materially to the strength and reliability of the joint.

Both riveted and bolted joints can be made in all sorts of different shapes and sizes but, broadly speaking, all such joints have a choice of three different ways of failing (Figure 10): (a) by shearing or breaking off the rivets themselves; (b) by tearing the rivets out of the plate (i.e. by 'bearing' or elongation of the holes); or (c) by breaking the material of one of the plates in tension between the rivets, like tearing off a postage stamp.

It is generally necessary to check the possibilities of failure by each of these three mechanisms by doing a suitable calculation. However, 'rules' for the design of riveted joints are laid down by organizations like Lloyds and the Board of Trade, and these are to be found in nearly all the engineering handbooks.

Welded joints

Welded joints of all kinds are very widely used in steelwork today, mainly because welding is generally cheaper than riveting and also because there is some increase in strength and saving in weight. In ships, too, the absence of rivet heads below the water-line reduces the resistance by a small amount.

Most sophisticated welding is electric arc welding. In this process the welder holds a metal rod, the welding rod, in his right hand by means of an insulated clamp. With his left hand he generally holds a mask or screen, furnished with very dark glass, through which he can safely watch the arc, which he 'strikes' and holds between the tip of the rod and the seam which he is making. At the usual 30–50 volts the arc is perhaps a quarter of an inch (7 mm) long and results in the transference of metal from the end of the welding rod to a little pool of molten steel which the welder coaxes along the joint. The result is, or should be, a continuous run or 'leg' of weld metal, about a quarter of an inch (7 mm) wide, which solidifies and bridges the joint. If a greater thickness of weld is needed, then the run must be repeated as many times as may be necessary.

If the weld has been properly made it is generally very strong and satisfactory, but any lack of skill or attention on the part of the welder is likely to result in defects, such as slag inclusions, which weaken the joint and are not readily seen by an inspector. It is also easy for a clumsy welder to overheat enough of the surrounding metal to cause serious distortions. This is especially the case where the work to be welded is heavy and thick; the welded engine-seatings in the pocket battleship *Graf Spee*, for instance, gave serious trouble from this cause.

In theory a welded joint in a tank or a ship should be completely watertight without further treatment, but this is seldom the case; in practice welded construction is likely to give more trouble than riveted work in this respect. A riveted lap joint is easily caulked by spreading the edges of the plates by means of a pneumatic chisel or caulking tool. This cannot be done with a welded joint, and the best way of dealing with the situation is to inject some kind of liquid sealing compound under pressure into the space between the two welds of a lap joint. All the same, I remember seeing much trouble in connection with the water-testing of compartments in welded warships.

Once upon a time I had the privilege of working for a few weeks as a riveter and also as a welder in one of the Royal Dockyards, and during this time I learnt various things which I do not think are in the text-books. Although closing a two-inch rivet in an armoured deck with a pneumatic hammer is both hard and noisy work, it is also curiously interesting, and most forms of riveting seemed to me to have at least some of the attraction of golf with the advantage of being more useful. A further sporting element was added by the operation of the inspection process; in those days we were paid at the rate of so much for every rivet closed, but five times so much was deducted for every rivet which was condemned by the inspector and had to be drilled out and replaced.

Riveting may not be heaven, but, by contrast, welding was certainly hell. Welding is amusing enough for the first hour or two – as I dare say hell may very possibly be – but after this the task of watching a hissing, flickering arc and a wretched little pool of molten metal becomes intolerably dull, and the dullness is not much relieved by the sparks and blobs of molten metal which find their way down one's neck and into one's shoes. After a very few days a feeling of boredom and bloody-mindedness settles in and it becomes very difficult to concentrate upon making a satisfactory weld.

Nowadays welds in tubing and in pressure vessels are made by automatic machines, which I suppose do not get bored, and so these welds are usually reliable. However, automatic welding is often impracticable in large structures like ships and bridges, where, in practice, the welding generally turns out to be very imperfect. Furthermore a welded joint provides little or no barrier to crack propagation, and this is one reason why so many large steel structures have failed catastrophically in recent years.

Creep

Homer knew that the first thing to do on getting your chariot out was to put the wheels on.

John Chadwick, *The Decipherment of Linear B* (Cambridge University Press, 1968)

The chariots of Mycenaean and archaic Greece had very light and flexible wheels, made from thin bent wood – willow or elm or cypress – usually with only four spokes (Figure 11). Such a construction was highly springy and resilient, and it seems to have enabled these vehicles to be galloped across the rough ground of the Greek hillsides, where a heavier and more rigid vehicle would

Figure 11. The Homeric chariot wheel was essentially flexible and made by bending quite thin wood. It could easily distort or 'creep' under any prolonged load.

have been useless. In fact, the rim of the wheel bent, rather like a bow, under the weight of the chariot, and, just as a bow must not be left strung for any length of time, so the weight must not be left on the wheels of a chariot. In the evening, therefore, one either tipped the vehicle vertically against a wall with the weight off the wheels, as Telemachus did in Book IV of the *Odyssey*, or else one took the wheels off altogether. Even on Mount Olympus the goddess Hebe had the morning chore of fitting the wheels to the chariot of grey-eyed Athene. With the much heavier wheels of later times such a procedure is less necessary and less practicable, although I understand that the wheels of the present Lord Mayor's coach are distinctly eccentric, presumably because the weight has been left on them for long periods.*

The distortion of bows and chariot wheels under prolonged

* This sort of thing is at the root of most of the stories about V.I.P.s being seasick when riding in state-coaches.

loading is due to what the engineer calls 'creep'. In elementary Hookean elasticity we assume, for simplicity, that if a material will sustain a stress at all, it will sustain it indefinitely, and also that the strains in a solid do not change with time, so long as the stress remains constant. In real materials neither of these assumptions is strictly true; nearly every substance will continue to extend or creep under a constant load with the passage of time.

The amount by which materials creep, however, varies a great deal. Among technological materials, wood and rope and concrete all creep very considerably and the effect has to be allowed for. Creep in textiles is one reason why our clothes go out of shape and the knees of trousers get baggy; it is, however, much

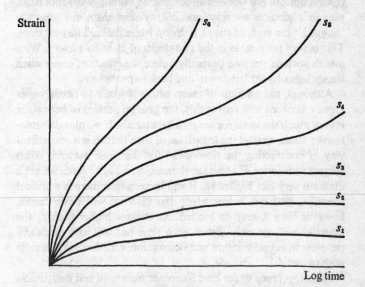

Figure 12. Typical time-creep curves for a material subjected to a series of constant stresses, $s_1, s_2, s_3 \ldots$ etc.

more pronounced in natural fibres, such as wool and cotton, than it is with the newer artificial fibres. This is why Terylene sails not only keep their shape but do not need to be carefully 'stretched' when new, as had to be done with cotton and flax sails.

Creep in metals is generally less pronounced than it is in non-metals, and, although steel creeps significantly at high stresses and when heated, the effect can often be neglected when one is dealing with light loads at ordinary temperatures.

Creep in any material causes the stress to be redistributed in a manner which is often beneficial, since the more highly stressed parts creep the most. This is why old shoes are more comfortable than new ones. Thus the strength of a joint may improve with age if the stress concentrations are diminished. Naturally, if the load on the joint is reversed, creep may have the opposite effect and the joint may be weakened.

The effect of the distortions caused by creep is particularly conspicuous in old wooden structures. In buildings the roof often sags in a picturesque way, and old wooden ships are generally 'hogged': the ends of the ship droop while the middle part rises. This is very noticeable in the gun-decks of H.M.S. *Victory*. With metals such as steel we generally notice the effects of creep when the springs of a car 'sit down' and have to be replaced.

Although the amount of creep which is likely to occur varies greatly between different solids, the general pattern of behaviour is very much the same for nearly all materials. If we plot deformation or strain against the logarithm of time (which is a convenient way of contracting the time-scale) for the same material when subject to a series of constant stresses, s_1, s_2, s_3 ... etc., we get a diagram very like Figure 12. It will be seen that there is a critical stress, s_3 perhaps, below which the material will never break, however long it may be loaded. At stresses higher than s_3 the material will not only distort with time but will also gradually progress to actual fracture and destruction, an effect we generally wish to avoid.

Soils, too, creep under load like other materials, and thus, unless we build upon rock or very hard ground, we need to watch the 'settlement' of foundations, which will usually need to be deeper for large buildings than for small. This is the reason for constructing large buildings on concrete 'rafts'. Note the subsidence of the foundations of the arches of Clare bridge in Plate 7.

Chapter 8   Soft materials and living structures

– or how to design a worm

> '*I'm very glad,*' *said Pooh happily,* '*that I thought
> of giving you a Useful Pot to put things in.*'
> '*I'm very glad,*' *said Piglet happily,* '*that I thought
> of giving you Something to put in a Useful Pot.*'

A. A. Milne, *Winnie-the-Pooh*

When Nature invented Something called 'life' she may have
looked around, a little anxiously, for a Useful Pot to put it in, for
life would not have prospered for long naked and unconfined. At
the time this planet presumably afforded rocks and sand, water
and an atmosphere of sorts, but it must have been rather short of
suitable materials for containers. Hard shells could be made from
minerals, but the advantages of a soft skin, particularly in the
earlier stages of evolution, seem to be overwhelming.

Physiologically, cell walls and other living membranes may
need to have a rather closely controlled permeability to certain
molecules but not to others. Mechanically, the function of these
membranes is often that of a rather flexible bag. They generally
need to be able to resist tension forces and to be able to stretch
very considerably without bursting or tearing. Also in most cases
skins and membranes have to be able to recover their original
lengths of their own accord when the force which has been
extending them is removed.* The strains to which present-day
living membranes can be extended safely and repeatedly varies a
good deal but may typically lie between 50 and 100 per cent. The
safe strain under working conditions for ordinary engineering
materials is generally less than 0·1 per cent, and so we might say
that biological tissues need to work elastically at strains which are
about a thousand times higher than those which ordinary technological solids can put up with.

*The mechanical problem is often much complicated by the association of
muscle tissue and other active devices for contraction, but we shall ignore
this for the present.

Not only does this enormous increase in the range of strain upset a number of the conventional engineer's preconceived ideas about elasticity and structures; it is also clear that strains of this magnitude cannot be furnished by solids of the crystalline or glassy type based on minerals or metals or other hard substances. It is therefore tempting, at least to the materials scientist, to suppose that living cells might have begun as droplets enclosed by the forces of surface tension. We must be quite clear, however, that it is very far from certain that this is what actually happened; what really did occur may have been something quite different – or at any rate considerably more complicated. What is certain is that some features of the elasticity of animal soft tissues resemble the behaviour of liquid surfaces and thus may possibly derive from them.

Surface tension

If we extend the surface of a liquid, so that it presents a larger area than before, we shall have to increase the number of molecules present at the surface. These extra molecules can only come from within the interior of the liquid and they will have to be dragged from the inside of the liquid to its surface against the forces which tend to keep them in the interior, which can be shown to be quite large. For this reason the creation of a new surface requires energy, and the surface also contains a tension which is a perfectly real force.* This is most easily seen in a drop of water or mercury, where the tension in the surface pulls the drop into a more or less spherical shape against the force of gravity.

When a drop hangs from the mouth of a tap, the weight of the water in the drop is being sustained by the tension in its surface. This phenomenon is the subject of a simple school experiment where one measures the surface tension of water and other liquids by counting the drops and weighing them.

Although the tension in a liquid surface is just as real as the tension in a piece of string, or any other solid, it differs from an elastic or Hookean tension in at least three important respects:

*The theory of surface tension was originally worked out, independently by Young and by Laplace, about 1805.

1. The tension force does not depend upon the strain or extension but is constant however far the surface is stretched.

2. Unlike a solid, the surface of a liquid can be extended, virtually indefinitely and to as large a strain as one cares to call for, without breaking.

3. The tension force does not depend upon the cross-sectional area but only upon the width of the surface. The surface tension is just the same in a deep or 'thick' liquid as it is in a shallow or 'thin' one.

Drops of liquid in air are of little use for biological purposes, because they soon fall to the ground; but droplets of one liquid floating within another liquid can continue to exist indefinitely and are of great importance both in biology and in technology. Systems of this kind are called 'emulsions' and are familiar in milk, in lubricants and in many kinds of paint.

Droplets are generally spherical and the volume of a sphere is as the cube of its radius, whereas the surface area of a sphere is as the square of the radius. Thus, if two similar droplets were to join up so as to make one droplet of twice the volume, there would be a considerable net reduction in surface area and so in surface energy. So there is an energy incentive for the drops in an emulsion to coalesce and for the system to segregate into two continuous liquids.

If we want the droplets to remain separate and not to coalesce, we have to arrange for them to repel each other. This is called 'stabilizing the emulsion' and is rather a complex process. One factor in stabilization is the provision of a suitable electrical charge on the surface of the drops – which is why emulsions are affected by electrolytes such as acids and alkalis. If stabilization has been done properly we have to do quite a lot of work to bring the droplets together – in spite of the saving in surface energy – which is why churning cream to make butter is hard work; Nature is rather good at stabilizing emulsions.

Although it does have some serious disadvantages, yet, as long as an animal is content to be very small and round, there is a good deal to be said for surface tension as a skin or membrane or container. For one thing, such a skin is very extensible and it is also self-healing; for another, the problem of reproduction is greatly simplified, since, if a droplet swells, it can break into two and become two droplets.

The behaviour of real soft tissues

As far as I know, no present-day cell wall operates simply by a straightforward surface tension mechanism; but many of them do behave in a way which is mechanically rather similar. One of the difficulties about simple surface tension is that the tension force is constant and cannot be increased by making the skin thicker; this limits the size of any container made in this way.

However, Nature is quite capable of producing a material which will have the characteristics of surface tension 'right through its thickness', so to speak. A slightly embarrassing example may be familiar to many people; when the dentist tells one to spit into his basin the resulting string or cord of saliva sometimes appears to be infinitely extensible and virtually unbreakable. What molecular mechanism is taking place is not at all clear, but the behaviour of such a material in terms of stress and strain is very much as in Figure 1.

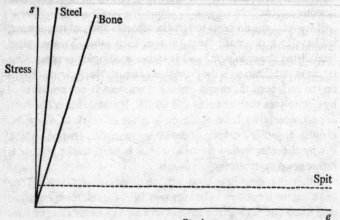

Figure 1. Stress-strain diagram for steel, bone and spit.

Most animal tissues are not as extensible as spit, but a very high proportion of them do show rather similar characteristics up to strains of 50 per cent or more. The urinary bladder in young people will stretch, more or less in this fashion, to around 100 per cent strain and that of dogs to about 200 per cent. As we mentioned in Chapter 3, my colleague Dr Julian Vincent has recently shown that, whereas the soft cuticle of male locusts and of virgin female locusts is content with a strain of something under 100 per cent, that of the pregnant female will stretch to an incredible 1,200 per cent – and still recover completely.

Although the stress-strain curve for most membranes and other soft tissues is not strictly horizontal, it is often very nearly so, at any rate up to the first 50 per cent or so of strain, and we may well consider what the consequences of this sort of elasticity are. In fact, any structure made from such a material must necessarily resemble one made from films of liquid under surface tension, and they are best observed by blowing soap bubbles next time you are in your bath.

The basic principle involved is that a material or membrane of this sort is essentially a constant-stress device – that is to say, it has only one stress to offer, and that one stress will operate in all directions. The only shape of shell or vessel or pressure container which is compatible with this condition is either a sphere or else a part of a sphere. This can be seen pretty clearly with soap-suds and in the froth on beer. If one should want to make an elongated animal from membranes of this sort, then the best thing to do seems to be to make it of a 'segmented' construction, like Figure

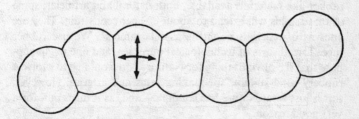

Figure 2.   A segmented animal. Stresses are equal in both directions in the surface.

2, and in fact this sort of thing is very common in worm-like creatures.

However successful this device may be for the cuticle of worms, it is of no use if what is wanted is a pipe or a tube, such as a blood-vessel. For pipes, as we saw in Chapter 6, the circumferential stress is ineluctably twice the axial stress, and this differential is just what a membrane of the sort we have been discussing cannot furnish. So it is necessary to have a material whose stress-strain curve slopes upwards after the fashion of Figure 3.

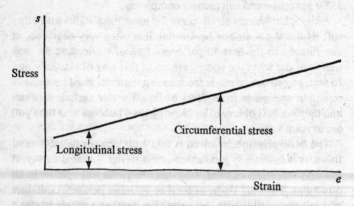

Stress

Circumferential stress

Longitudinal stress

Strain

Figure 3. To make the skin of a cylindrical container the stress-strain curve of the membrane must slope upwards so as to afford a circumferential stress which is twice the longitudinal stress.

The most obvious kind of highly extensible solid which fulfils this condition is rubber, and there is nowadays a wide range of rubber-like materials available, both natural and artificial; some of these solids will extend to about 800 per cent strain. They are known to materials scientists as 'elastomers'. We use rubber tubes for all sorts of technological purposes, and one might suppose that the obvious thing for Nature to do would be to evolve a rubbery solid suitable for making veins and arteries. However, this is just what Nature has *not* done – and, as it turns out, for a very good reason.

Materials of the rubbery kind have a stress-strain curve which has a very characteristic 'sigmoid' or 'S' shape (Figure 4). Accord-

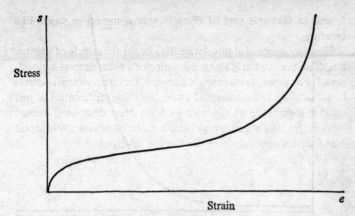

Figure 4.   Stress-strain curve for typical rubber.

ing to my rather shaky mathematics, one can show that if we make a tube or a cylinder from such a material and then inflate it, by means of an internal pressure, so as to involve a circumferential strain of 50 per cent or more, then the inflation or swelling process will become unstable, and the tube will bulge out, like a snake which has eaten a football, into a spherical protrusion which a doctor would describe as an 'aneurism'. Since one can easily produce this result experimentally by blowing up an ordinary child's cylindrical rubber balloon (Plate 3), my mathematics are probably right.

Since veins and arteries do, in fact, generally operate at strains around 50 per cent, and since, as any doctor will tell you, one of the conditions it is most desirable to avoid in blood-vessels is the production of aneurisms, any sort of rubbery elasticity is quite unsuitable for most of our internal membranes; and in fact it is comparatively rare in animal tissues.

When we do the mathematics it turns out that the only sort of elasticity which is completely stable under fluid pressures at high strains is that which is represented by Figure 5. With minor variations, this shape of stress-strain curve is very common indeed for animal tissue and particularly for membranes. You can feel that this is so if you pull on the lobe of your ear.

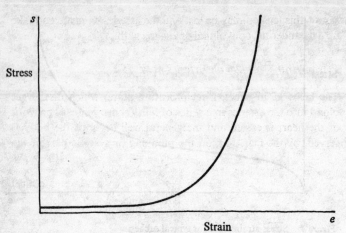

Figure 5. Stress-strain curve for typical animal tissue.

It will be noticed that Figure 5 seems to beg the question of whether the stress-strain curve for such materials really passes through the origin (the point of zero stress and strain) or whether there is still a finite tension in the material when there is no strain – a state of affairs which is no doubt calculated to shock the souls of engineers brought up on Hookean materials like steel. As far as one can see, however, in the living body there seems to be nothing really corresponding to an 'origin': that is, there is, apparently, no real position of zero stress and strain (as there would be in any structure made from, say, soap-films). The arteries, at any rate, are permanently in tension within the body, and, if they are dissected out of a living or a freshly dead animal, they will shorten fairly noticeably.

As we shall see in the next section, this tension is perhaps an additional device to counteract any tendency of the artery to alter in length as the blood-pressure changes, or else it may represent a belated attempt to equalize the longitudinal and circumferential stresses within the artery wall – in other words an attempt to get back to the conditions of surface tension which may have existed in the dim past. When people are subjected to severe and prolonged vibration – for instance in the case of foresters using chain saws – this tension may be lost and the arteries elongate and take up a meandering, convoluted or zig-zag path.

Poisson's ratio – or how our arteries work

The heart is, in effect, a reciprocating pump which discharges blood into the arteries in a series of fairly sharp pulses. The work of the heart is eased, and the general well-being of the body is served, by the fact that, on the pumping or systolic part of the

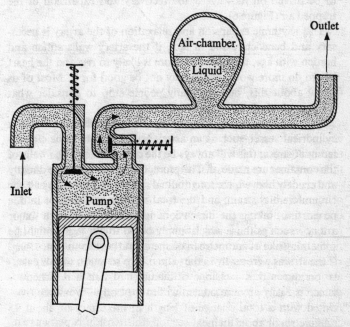

Figure 6.   The elastic expansion of the aorta and the arteries performs the same function in smoothing the fluctuations of blood-pressure as does the air-chamber attached to an engineer's reciprocating pump.

cardiac cycle, much of the excess of high-pressure blood is accommodated by the elastic expansion of the aorta and of the larger arteries; this has the effect of smoothing the fluctuations in pressure and generally facilitating the circulation of the blood. In fact the elasticity of the arteries does much the same job as the air-bottle affair which engineers often attach to mechanical reciprocating pumps. In this simple device the surge of pressure which accompanies the discharge stroke of the pump piston is smoothed out by arranging for the liquid which is being pumped temporarily to compress a supply of air which is trapped in a suitable bulb or container. When the valve of the pump shuts at the end of its stroke (as that of the heart does in diastole), the liquid continues to be driven on its way by the recovery and expansion of the trapped air (Figure 6).

This rhythmic expansion and relaxation of the artery is necessary and beneficial; and in fact, if the artery walls stiffen and harden with age, the blood-pressure is likely to rise and the heart has to do more work, which may not be good for it. Most of us know about this, but not many people stop to consider what happens about the strains in the artery walls.

As we calculated in Chapter 6, the longitudinal stress in a cylindrical vessel, such as an artery wall, is just half the circumferential stress; this will always be the case, whatever the walls of the container are made of. Therefore, if Hooke's law were directly and crudely obeyed, the longitudinal strain would also be half the circumferential strain, and the total extensions would be in due proportion, taking the dimensions into account. Now a major artery – such as those which supply blood to our legs – might be something like a centimetre in diameter and perhaps a metre long. If the strains were really in the ratio of two to one, a simple calculation shows that a change of diameter of half a millimetre – which is easily accommodated within the body – would be associated with a total change of length in the artery of about 25 millimetres or about an inch.

It is obvious that a change of length of this magnitude, occurring seventy times a minute, cannot and does not take place. If this sort of thing really happened, our bodies just would not work. To take an extreme case, one has only to imagine such a thing happening in the blood-vessels which serve the brain.

Fortunately, in real life, the lengthwise strains and extensions in pressurized tubes of all sorts and kinds are very much less than one might have anticipated or feared from this rather too simple argument. That this is so is due to something called 'Poisson's ratio'.

If you stretch a rubber band it gets very noticeably thinner; and much the same thing happens with all solids, although with most materials the effect is less conspicuous. Contrariwise, if you shorten a material by compressing it, it will bulge out sideways. Both of these are elastic effects and they disappear when the loads are taken off.

The reason why we do not notice these lateral movements in things like steel and bone is that both the longitudinal and the transverse strains are so small; but the effects are there all the same. The fact that this happens in all solids and that such behaviour is significant in practical elasticity was first observed by the Frenchman S. D. Poisson (1781–1840). Although he was born into rather dramatic poverty and got very little formal education before he was fifteen, Poisson was made an Academician – one of the highest honours France has to offer – at the age of thirty-one for his work on elasticity.

As we said in Chapter 3, Hooke's law says that

$$\text{Young's modulus} = E = \frac{\text{stress}}{\text{strain}} = \frac{s}{e}$$

Thus, if we apply a tensile stress, s_1, to a flat plate, the material will be elongated or stretched elastically so that there will be a tensile strain in the direction in which we are pulling of

$$e_1 = \frac{s_1}{E}$$

However, the material will also be *contracted* sideways (i.e. at right angles to s_1) by some other strain which we may call e_2. Poisson found that, for any given material, the ratio of e_2 to e_1 is constant, and this ratio is what we now call 'Poisson's ratio'. We shall use the symbol q in this book. Thus, for a given material subject to a simple uniaxial tension stress s_1,

$$q = \frac{e_2}{e_1} = \text{Poisson's ratio*}$$

e_1, the strain in the direction of s_1, is often called the 'primary strain'; the strain caused by s_1 at right angles to itself, so to speak, is called the 'secondary strain' (Figure 7).

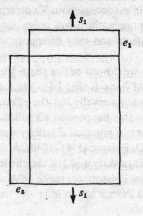

Figure 7. When a solid is stretched by a tensile stress s_1, it extends in the direction of s_1 by a primary strain e_1, but it also *contracts* laterally by a secondary strain e_2.

$$\text{Poisson's ratio} = q = \frac{e_2}{e_1}$$

From what we have just been saying,

$$e_2 = q \cdot e_1$$

and since $e_1 = s_1/E$ (which is Hooke's law)

then $$e_2 = q \cdot s_1/E$$

Thus if we know q and E we can calculate both the primary and the secondary strains.

*Since in all such cases, e_2 is always of opposite sign to e_1, q or Poisson's ratio ought always to be negative, and it should therefore carry a minus sign about with it. However, we choose to forget about this and omit the minus sign; this is compensated for by putting a minus sign in the sums, such as those we are now doing.

For engineering materials like metals and stone and concrete, q nearly always lies between $\frac{1}{4}$ and $\frac{1}{3}$. The values for Poisson's ratio for biological solids are generally higher than this and are often around $\frac{1}{2}$. Teachers of elementary elasticity will tell you that Poisson's ratio cannot have a higher value than $\frac{1}{2}$ – otherwise various naughty and inadmissible things would happen. This is only partly true, and the values for some biological materials can sometimes be very high indeed, often well over unity.* The experimental value for the Poisson's ratio for my tummy, measured recently by me in my bath, is about $1\cdot0$ (see the footnote on p. 162).

Thus, as we have said, the effect of Poisson's ratio is that, if we pull upon a piece of material, such as a membrane or an artery wall, in one direction it will get longer in that direction, but it will contract, or get shorter, in the direction at right angles. So if *two* tensions are applied, at right angles to each other, the effects will be additive and the strains will be *less* than we should expect if either of the stresses were applied separately.

For two simultaneous stresses, s_1 and s_2, the total strain in the direction of s_1 will be

$$e_1 = \frac{(s_1 - q s_2)}{E}$$

and the total strain in the direction of s_2 will be

$$e_2 = \frac{(s_2 - q s_1)}{E}$$

Harking back to Chapter 6, the consequence of the existence of Poisson's ratio is that the longitudinal or lengthwise strain in the wall of a tubular pressure vessel which obeys Hooke's law is

$$e_2 = \frac{rp}{2tE}(1 - 2q)$$

where r = radius, p = pressure, and t = wall thickness.

It follows that the longitudinal elastic extension of a tube is much

*To save indignant elasticians from the trouble of unnecessary correspondence, I do know about the energy changes involved. These anomalies have a rational explanation.

less than one might expect; for a Hookean material with a Poisson's ratio of $\frac{1}{2}$ there will be no movement at all. In fact, as we have seen, the artery walls do not obey Hooke's law, and it is also probable that their Poisson's ratio is higher than $\frac{1}{2}$; possibly these two effects offset each other, because experimentally very little lengthwise movement is observed.* No doubt the fact that arteries are permanently stretched within the body is a precaution against any residual longitudinal strain.

The effects of Poisson's ratio are probably of very great importance in animal tissues; but they are also significant in engineering and the matter is continually cropping up in all sorts of connections.

It should perhaps be added that, whereas the aorta and the principal arteries expand and contract elastically with each beat of the heart, in the manner we have just been discussing, the state of affairs with the smaller arteries is usually rather different. The walls of these lesser vessels are provided with muscular tissue which can increase their effective stiffness and so, by restricting the diameter, is able to control the amount of blood which is able to pass to any particular region of the body. In this way the local distribution of the blood-supply is adjusted.

Safety – or the toughness of animals

Animals quite often break their bones and they sometimes tear their tendons, neither of which have the sort of elasticity we have

---

* *Note for bio-elasticians.* This Hookean analysis is simplistic. For a non-Hookean system, where the *tangent* moduli are E_1 and E_2, then, approximately, the change of longitudinal strain is zero when

$$\frac{E_1}{E_2} = 2q$$

Although most soft tissues preserve approximately constant volume – that is, they seem to have a true Poisson's ratio around 0·5 – most membranes choose to deform in plane strain, that is to say, they do not get thinner when they are stretched, and so they show an apparent Poisson's ratio of about 1·0 – like my tummy. This fits with a value of E_1/E_2 of around 2·0, which is likely enough. But *why* does the membrane not get thinner when it is strained? See, for instance, E. A. Evans, *Proc. Int. Conf. on Comparative Physiology* (1974; North Holland Publishing Company).

been discussing; but it is very noticeable that the mechanical fracture of soft tissues seems to be rare. There are several reasons for this. Being so soft, skin and flesh can sometimes evade the effects of a blow by deflecting and escape with a bruise. The question of stress concentrations, however, seems to be more interesting, since the majority of soft animal tissues appear to be almost immune from this major cause of engineering catastrophes. For this reason the need for a factor of safety is much reduced, and thus the structural efficiency, that is, the load which the structure carries in proportion to its weight, may be quite high.

This immunity is not just a matter of being soft and having a low Young's modulus. Rubber is indeed soft and has quite a low modulus, yet many of us can remember, as children, having taken our blown-up rubber balloons out into the garden where they very soon burst with a bang on encountering the prickles of the first rose-bush. As children, we did not realize that, owing to the stress concentration and to the low work of fracture of rubber, a crack spreads very rapidly from a pin-hole in stretched rubber, and it is rather doubtful if our tears would have been much abated if we had. However, the membrane of a bat's wing, for example, although it is much stretched in flight, does not seem to behave in this way. If the wing does get punctured, the tear seldom spreads and the injury soon heals, although the bat may be using its wings continually.

The explanation lies, I think, in the very different elasticities and works of fracture of rubber and of animal membranes. There are at present virtually no data available about the works of fracture of biological soft tissues; but the shapes of the stress-strain curves are in most cases pretty well known, and this latter factor does seem to have a big influence upon the probability of fracture.

The shell-membrane of an egg seems to afford an interesting example – this is the membrane which you encounter at breakfast, just inside the shell of your boiled egg. It is one of the few biological membranes which obey Hooke's law, in this case up to its breaking strain of about 24 per cent. A simple but slightly messy experiment with a raw egg will show that egg membranes tear very easily. This is, of course, what they are there for, since the first thing the chick has to do is to get out of its egg, which it does by pecking with its beak. Incidentally the egg-shell itself, with its rounded domed shape, is difficult to break from the outside but easy to break from inside.

Egg membranes are rather exceptional, in that they exist in order to be broken after they have served their purpose of conserving the moisture in the egg and keeping out infection; as we have said, they possess a special sort of elasticity, very possibly for this reason. However, the great majority of soft tissues have an elasticity which is quite different and is very much like Figure 5; functionally, most of these tissues need to be tough. Although all the scientific reasons are not perfectly clear, it does seem that, pragmatically, materials with this type of stress-strain curve are extremely difficult to tear. One reason is, perhaps, that the strain energy stored under such a curve – and therefore available to propagate fracture (Chapter 5) is minimized.*

As we have said, a very high proportion of animal tissues behave, elastically, much in the manner of Figure 5. I must confess that, when this information first dawned on me, it seemed to me to indicate an eccentricity or quirk on the part of Nature, who, poor thing, did not know any better, not having had the benefit of an engineering education. After a good deal of rather blundering research into the elementary mathematics of the problem it is now beginning to dawn on me that, if one has need of a structural system which will work reliably at really high strains, this is the *only* sort of elasticity which will serve. In fact, the achievement of this kind of stress-strain curve in animal materials represents a really essential condition for the evolution and continued existence of the higher forms of life. Biologists, please note.

The constitution of soft tissues

Perhaps partly for these reasons the molecular structure of animal tissue does not often resemble that of rubber or artificial plastics. Most of these natural materials are highly complex, and in many cases they are of a composite nature, with at least two compo-

---

*The shape of the stress-strain curve for most animal tissues – such as skin – is very much like that of a knitted fabric, which it is almost impossible to tear.

nents; that is to say, they have a continuous phase or matrix which is reinforced by means of strong fibres or filaments of another substance. In a good many animals this continuous phase or matrix contains a material called 'elastin', which has a very low modulus and a stress-strain curve something like Figure 8. In

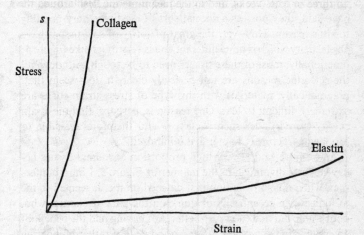

Figure 8.   Approximate stress-strain curves for elastin and collagen.

other words elastin is only about one stage removed, elastically, from a surface tension material. The elastin is, however, reinforced by an arrangement of bent and zig-zagged fibres of collagen (Plate 4), a protein, very much the same as tendon, which has a high modulus and a nearly Hookean behaviour. Because the reinforcing fibres are so much convoluted, when the material is in its resting or low-strain condition they contribute very little to its resistance to extension, and the initial elastic behaviour is pretty well that of the elastin. However, as the composite tissue stretches the collagen fibres begin to come taut; thus in the extended state the modulus of the material is that of the collagen, which more or less accounts for Figure 5.

The role of the collagen fibres is not merely to stiffen the tissue at high strains; they also seem to contribute very much to its toughness. When living tissue is cut, either accidentally or surgically, in the first stage of the healing process the collagen fibres are re-absorbed and disappear, temporarily, for a considerable distance around the wound. It is only after the gap has been filled and bridged by elastin that the collagen fibres are re-formed and the full strength of the tissue is restored. This process may take up to three or four weeks, and in the meantime the flesh around the

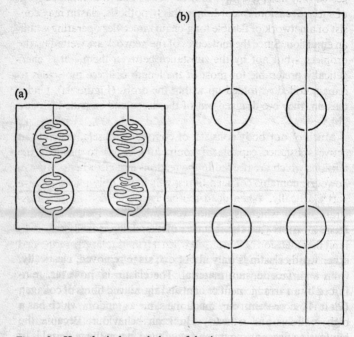

Figure 9.   Hypothetical morphology of elastin.
(a) Resting or unextended state. Chain molecules folded or mainly folded within droplets.
(b) Extended state. Chain molecules pulled out of droplets.

wound has an almost negligibly low work of fracture. It is for this reason that, if a surgical wound has to be reopened within two or three weeks of the original operation, it may be difficult to get the new stitches to hold.

Collagen exists in various forms, but it may consist of twisted strings or ropes of protein molecules, and its resistance to extension is basically due to the need to stretch the bonds between the atoms in the molecules: that is to say, it is a Hookean material much like nylon or steel. Why then does elastin behave as it does, almost like surface tension? The short answer is that nobody really knows, but Professors Weis-Fogh and Andersen have suggested that this behaviour may in fact be due to a modified form of surface tension. According to this hypothesis, elastin may consist of a network of flexible long-chain molecules operating within an emulsion. Since the molecules of the network are wetted by the droplets – but not by the substance between them – it is energetically preferable for most of the length of these molecules to remain coiled or folded up within the drops (Figure 9a). Under tension, they are dragged out of the drops and extended (Figure 9b).*

Much of our body consists, of course, of muscle, which is an active substance capable of contracting so as to produce the tensions which are needed in the tendons and elsewhere. Muscle, however, contains collagen fibres, which can only play a passive part elastically. When dead muscle is stretched it has a stress-strain curve which is, again, very much like Figure 5, and it seems possible that the function of the collagen in muscle is to limit the extension of the muscle when it is in its relaxed or extended state: in other words it acts as a sort of safety-stop.

As we have said, another purpose of collagen fibres in flesh is to put up the work of fracture. This is a good thing for the animal, but it is inconvenient for the people who want to eat its flesh. In other words, it is collagen which makes meat tough. Nature, however, does not seem to be on the side of the vegetarians, because she has arranged, in her wisdom, that collagen should break down to gelatin – a substance of low strength when wet – at a somewhat lower temperature than that which elastin or muscle can withstand. The process of cooking meat therefore consists in converting most of the collagen fibres into gelatin (which is jelly or glue) by roasting or frying or boiling. It is science of this kind which restores one's faith in the beneficence of Providence.

* Since this was written, Dr J. M. Gosline has put forward an alternative hypothesis to account for the behaviour of elastin.

Part Three

Compression and bending structures

Chapter 9   Walls, arches and dams

– or cloud-capp'd towers and the stability of masonry

> *What are you able to build with your blocks?*
> *Castles and palaces, temples and docks.*
>
> R. L. Stevenson, *A Child's Garden of Verses*

As we have seen, unless one is as clever as Nature is, the whole business of making tension structures is set about with difficulties, complications and treacherous traps for the unwary. This is especially the case when we want to make a structure from more than one piece of material, so that we are faced with the problem of preventing it from coming apart at the joints. For these reasons our ancestors generally avoided tension structures as far as they could and tried to use constructions in which everything was in compression.

Much the oldest and the most satisfactory way of doing this is to use masonry. As a matter of fact the immense success of masonry buildings has really been due to two factors. The first is the obvious one about avoiding tension stresses, especially in the joints; the second reason may be less obvious. It is that the nature of the design problem in large masonry buildings is peculiarly adapted to the limitations of the pre-scientific mind.

Out of all the different kinds of structures which might be made, the masonry building is, as we shall see, the *only* one in which a blind reliance on traditional proportions will not automatically lead to disaster. This is why, historically, masonry buildings were by far the largest and most imposing of the works of man. The desire to build cloud-capp'd towers and solemn temples goes far back into history and indeed into prehistory. There is a quotation from Genesis about the Tower of Babel at the head of Chapter 1. It may be remembered that this was a project to build 'a tower with its top in the heavens'. However, I do not think any theologian has ever inquired to what height such a tower could really have been built.

Nearly all the load upon the walls would have been due to the effect of their own weight, and one way of looking at the problem is to calculate the direct compressive stress which would be caused near the bottom of the tower by the vertical dead weight of the masonry. A limit will be set to the height of the structure when the bricks begin to be crushed by the superincumbent weight.

Now brick* and stone weigh about 120 lb. per cubic foot (2,000 kg/m³), and the crushing strength of these materials is generally rather better than 6,000 p.s.i. or 40 MN/m². Elementary arithmetic shows that a tower with parallel walls could have been built to a height of 7,000 feet or 2 kilometres before the bricks at the bottom would be crushed. However, by making the walls taper towards the top, it could have been made much higher still; this is more or less how mountains work. Mount Everest is 29,028 feet or about 8 kilometres high and shows no signs of collapsing. Thus a simple tower, preferably with a broad base and tapered towards the top, could well have been built to such a height that the men of Shinar would have run short of oxygen and had difficulty in breathing before the brick walls were crushed beneath their own dead weight.

Although there is nothing very much wrong with this sum, in fact even the most ambitious towers have never been built to anything remotely approaching that kind of height. The tallest 'building' which actually exists today is probably the New York Trade Center, which is only about 1,350 feet or 400 metres high; and this, like other skyscrapers, could be said to be cheating, since its structure is made of steel. The Great Pyramid and the highest cathedral spires reach a little more than 500 feet or 150 metres, but very few other masonry buildings are more than half so tall; the great majority are much lower still.

Therefore the compressive stresses in everyday masonry due to its own vertical dead weight are very small indeed. In general they are seldom more than a hundredth part of the crushing strength of the stone, and so this factor is not, in practice, a

---

* Note that Genesis 11 specifically says 'let us make bricks and bake them hard'. There was no question of using cheap mud bricks as the Egyptians did. This seems to be an early example of the Concorde syndrome.

limitation upon the height or the strength of buildings. However – to be biblical again – the Tower of Siloam, which was probably not particularly high, fell and killed eighteen people, and it is notorious that in spite of the confidence of builders and architects, walls and buildings do fall down unexpectedly. They have been doing so for a very long time and they still sometimes do so today. Since masonry is heavy, people often get killed.

If walls do not collapse because of the direct crushing stress upon the material, why do they fall down? Once again, we can learn from what children do. When we were very young, most of us played with 'bricks', and about the first thing we did was to build a tower by piling one brick upon another rather erratically. Usually, when the tower had reached a modest height, it fell down. Even the child knew perfectly well, although he could not have expressed the idea in scientific words, that there was no question of the bricks being crushed under a compressive stress. The actual stress in the bricks was negligible; what happened was that the pile of bricks tipped up and fell over because the tower was not straight and vertical. In other words the failure was due to lack of stability and not to lack of strength. Although this distinction soon becomes evident to young children, it is not always clear to builders and architects. For the same reason the reflections of art historians who write about cathedrals and other buildings are apt to make rather distressing reading.

Thrust lines and the stability of walls

> How reverend is the face of this tall pile,
> Whose ancient pillars rear their marble heads,
> To bear aloft its arched and ponderous roof,
> By its own weight made steadfast and immoveable,
> Looking tranquillity. It strikes an awe
> And terror on my aching sight.

William Congreve, *The Mourning Bride*

There was only one culture in Queen Anne's time, and there is very little doubt that Congreve (1670–1729) talked and drank with Vanbrugh, who wrote plays and designed Blenheim Palace, and also with Sir Christopher Wren himself. To all these people it was perfectly clear – in a general kind of way – that what kept a building from tipping up and collapsing was not so much the strength of the stones and mortar as the weight of the material, acting in the right places.

However, it is one thing to be aware of this in a general kind of way and another to understand what is happening in detail and to be able to predict just when a building is safe and when it is not. In order to get a proper scientific understanding of the behaviour of masonry it is necessary to treat it as an elastic material; that is to say, one must take into account the fact that the stones deflect when they are loaded and that they obey Hooke's law. It is also a considerable help, though perhaps not absolutely essential, to make use of the concepts of stress and strain.

At first sight it does, of course, seem improbable that solid brick and stone should deflect to any significant extent under the loads which occur in a building. In fact for at least a century after Hooke's time the common-sense view prevailed, and builders and architects and engineers persisted in ignoring Hooke's law and treating masonry as if it were perfectly rigid. In consequence, their buildings sometimes fell down because they got their sums wrong.

As a matter of fact the Young's moduli of brick and stone are not particularly high, and, as one can see from the bent pillars in Salisbury Cathedral (Plate 1), the elastic movements in masonry are by no means so tiny as one might suppose. Even in an ordinary small house the walls are likely to be shortened or compressed elastically, in the vertical direction, by something like a millimetre under their own weight. In a large building the movements are naturally much greater. Incidentally, when the house is shaken by the wind during a gale, you are not imagining the effect; the house *is* being shaken by the wind. The top of the Empire State building sways about two feet during a storm.*

* Of the abbey church of Saint Denis, in France, during the twelfth century, we read '. . . such a force of contrary gales hurled itself against the aforesaid arches, not supported by scaffolding nor resting on any props, that they threatened baneful ruin at any moment, miserably trembling and, as it were, swaying hither and thither.' (I am indebted to Professor Heyman for this reference.)

The modern analysis of masonry structures is based upon simple Hookean elasticity and also upon four assumptions, all of which turn out to be justified by practical experience. These are:

1. That the compressive stresses are so small that the material will not be broken by crushing. We have already discussed why this is so.

2. That, owing to the use of mortar or cement, the fit between the joints is so good that the compressive forces will be transmitted over the whole area of the joint and not just at a few high spots.

3. That the friction in the joints is so high that failure will not happen because of bricks or stones sliding over each other. In fact no sliding movements at all will take place before the structure collapses.

4. That the joints have no useful tensile strength. Even if, by chance, the mortar does have some strength in tension, this cannot be relied upon and must be neglected.

Thus the function of the mortar is not to 'glue' the bricks or stones together but simply to transmit the compressive load more evenly.

As far as I know, the first person to take the elastic deformations of masonry into account was Thomas Young. Young considered what would happen in a rectangular block of masonry, such as a piece of a wall, when it had to carry a vertical compressive load, P, let us say. In what follows I have simplified Young's arguments by translating them into the language of stress and strain, which of course was not available in his time.

As long as P acts symmetrically along the centre-line, that is, down the middle of the wall, then the masonry will be compressed uniformly, and, because of Mr Hooke, the corresponding distribution of compressive stress across the thickness of the wall will also be uniform (Figure 1).

Suppose, now, that the vertical load, P, becomes a little eccentric, that is to say, it no longer acts exactly along the centre-line; then the compressive stress can no longer be spread evenly but must be higher on one side than on the other so as to react properly against the load and keep it in balance. If the material obeys Hooke's law, then Young showed that the stress will be distributed linearly and the stress-distribution diagram will look like Figure 2.

So far the mortar in the joint is quite happy because the whole width of the joint is still safely in compression. However, if the

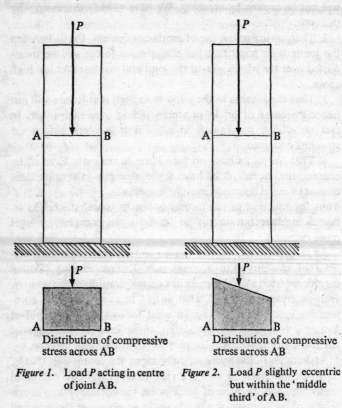

Distribution of compressive stress across AB

Distribution of compressive stress across AB

Figure 1. Load *P* acting in centre of joint A B.

Figure 2. Load *P* slightly eccentric but within the 'middle third' of A B.

position of the load is displaced still further from the centre – in fact to the edge of what is called the 'middle third' of the wall – then a situation like Figure 3 will arise where the load distribution is now triangular and the compressive stress at the outside edge of the joint is zero.

This, in itself, does not matter too much, but it must be becoming clear to the percipient mind that something is about to happen. In fact, if the load is now displaced a little further outwards, something *will* happen: that is to say, a situation like Figure 4.

The stress at the opposite surface of the wall has now changed from compression to tension. We said, however, that mortar

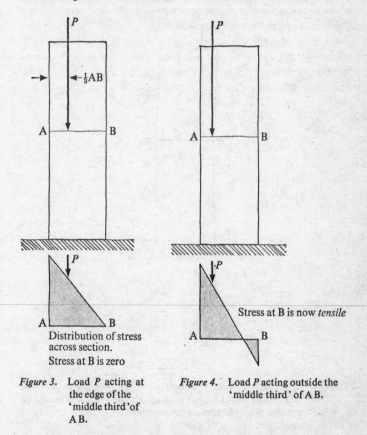

Figure 3. Load *P* acting at the edge of the 'middle third' of A B.

Figure 4. Load *P* acting outside the 'middle third' of A B.

cannot be trusted to take tension, and this is generally only too true. What one would expect to happen usually does happen: the joint cracks. Of course it is a bad thing for walls to crack, and it should not be allowed to happen in well-regulated buildings, but it does not necessarily follow that the wall is going to fall down immediately. What is likely to occur in real life is simply that the crack will gape a bit but the wall will continue to stand up, resting on the parts which are still in contact (Figure 5).

All this savours somewhat of living dangerously, and one of these days the line of the thrust may stray *outside* the surface of

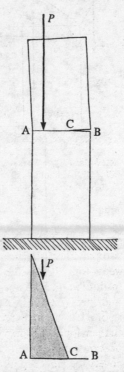

Figure 5.   What really happens as a result of the condition drawn in Figure 4. The joint cracks from B to C, and the load is now carried over the area A C – effectively a narrower wall.

the wall, when, as a little thought will show, since no tension forces are available, one or more of the joints will hinge about its outside edge and the wall will tip up and fall down (Figure 6). It really will.

At the time when he came to these conclusions, that is, about

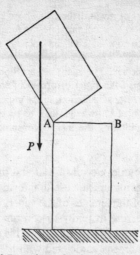

Figure 6. When load *P* acts *beyond* A, i.e. outside the surface boundary of the wall, the wall will hinge about A, tip up and fall.

1802, Young, a rising man of twenty-nine, was appointed to the chair of Natural Philosophy at the Royal Institution in London. His colleague, and in some sense his rival, was Humphry Davy, who was made Professor of Chemistry in the same year at the improbable age of twenty-four. It was the custom of the professors of the Royal Institution, then as now, to deliver series of lectures to popular audiences. In those days, however, these lectures had very much of a television character, and the Institution relied heavily upon them for both money and publicity.

Young took his educational mission seriously, and, filled with the enthusiasm of discovery, he launched into a series of lectures about the elasticity of various kinds of structures, with many useful and novel observations on the behaviour of walls and arches.

The audience at Albemarle Street in those days was a fashionable one and is said to have consisted largely of 'silly women and dilettante philosophers'. Young by no means neglected the feminine part of his audience, and he remarked in his opening lecture:

A considerable part of my audience, to whose information it will be my particular ambition to accommodate my lectures, consists of that sex which, by the custom of civilized society, is in some measure exempted from the more laborious duties which occupy the time and attention of the other sex. The many leisure hours which are at the command of females in the superior orders of society may surely be appropriated, with greater satisfaction, to the improvement of the mind and to the acquisition of knowledge than to such amusements as are only designed for facilitating the insipid consumption of superfluous time ...

However, fortune does not always attend those who, however earnestly, strive to communicate useful information, and one may suspect that some of the females of the superior orders of society slipped away, preferring insipidly to consume their superfluous time. In any case Davy, who exhibited in his own lectures some of the exciting phenomena associated with the new electric fluid, together with a range of colourful chemical experiments, was a pushing young particle with what we should now call a television personality. Davy was also remarkably good-looking, and young women flocked to his lectures for reasons which were not always strictly academic; 'those eyes', one of them was heard to say, 'were made for something besides poring over crucibles.' The result, in box-office terms, could not be in doubt, and we are told that

Dr Young, whose profound knowledge of the subjects he taught no one will venture to question, lectured in the same theatre and to an audience similarly constituted to that which was attracted to Davy, but he found the number of his attendants diminish daily and for no other reason than that he adopted too severe and too didactic a style.

This kind of failure might not have mattered too much if Young could have attracted the interest and support of practical engineers. However, the engineering profession at that time was led, and frequently dominated, by the great Thomas Telford (1757–1834), whose views, as we have seen, were severely pragmatic and anti-theoretical. In consequence Young resigned his chair almost immediately and returned to his medical practice.* The development of elasticity passed, for many years, to France, where, at this time, Napoleon was actively encouraging the study of structural theory.

The theory about elastic compression, the 'middle third' and instability which so bored the fashionable females at Young's lectures does really tell us practically all we need to know about the behaviour of joints in masonry, *provided* that we also know the position at which the weight can be considered as acting. In other words, how eccentric is the load?

This is best determined by means of what is called the 'thrust line', that is to say, a line passing down the wall of a building

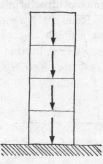

Figure 7. For the simplest symmetrical case, the 'thrust line' passes down the centre of a wall.

from the top to the bottom which defines the position at which the vertical thrust can be considered as acting in each successive joint. The thrust line is a French invention and seems to have been first thought of by Coulomb (1736–1806).

For a very simple symmetrical wall or column or pillar, such as Figure 7, the thrust line will clearly pass down the middle of the wall and so there is really no problem. However, in a building with any pretension to sophistication there is most likely to be at least one oblique force arising from the sideways thrust of the

* Davy remained at the Royal Institution and prospered. He became Sir Humphry and President of the Royal Society. He is said to have been offered a bishopric if he would take Holy Orders. As a great man who had risen from humble beginnings he behaved rather badly to a coalminer called George Stephenson but rather well to a blacksmith's son called Michael Faraday.

roof members, from archways or vaultings or from various other forms of asymmetrical construction. In such a case the thrust line will no longer pass neatly down the middle of the wall but will be displaced to one side, frequently into a curved path such as Figure 8.*

If, on plotting the thrust line, we find that it is in danger of reaching the surface of the wall at any point, then we shall clearly

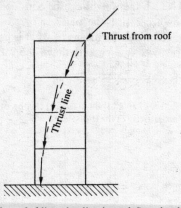

Figure 8. The effect of oblique loading is to deflect the thrust line in this kind of way.

have to think again, and think hard, because there is a good chance that a building designed like that will fall down.

One of the things we can do, and it may well be one of the most effective, is to add weight to the top of the wall. What then happens can be represented diagrammatically by Figure 9. Contrary to what one might suppose, weight at the top is likely to make a wall more, and not less, stable and will bring an erring thrust line back, more or less, to where it ought to be.

One way to do this is simply to build the wall to a greater

*That this is so can be checked by applying the parallelogram of forces (whose acquaintance can be renewed in the pages of elementary text-books on mechanics) at each level in the wall. The parallelogram of forces is supposed to have been invented by Simon Stevin in 1586. The absence of the concept of the resolution of forces is one reason why it is impossible that either ancient or medieval architects could have designed their buildings in the modern way.

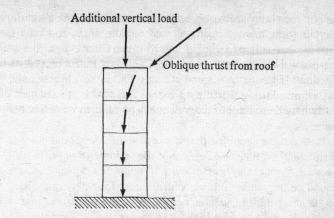

Figure 9. The effect of an additional load at the top of the wall is to *reduce* the eccentricity of the thrust line.

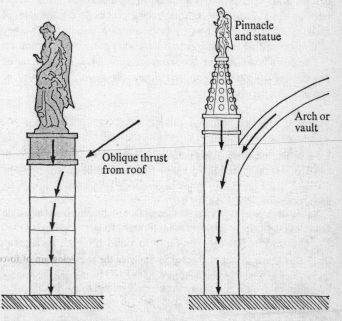

Figure 10. This can be done by adding top weight in the form of pinnacle, statues etc.

height than appears to be really necessary, and, in addition, anything like heavy balustrades and copings are a good thing. If it is that sort of building and you can afford it, a line of statues will always help (Figure 10). This is the structural justification for the pinnacles and statuary on Gothic churches and cathedrals. They are really up there to say 'boo' to the functionalists and to all the dreary people who bleat too much about 'efficiency'.

It used to be supposed that it was absolutely essential that the thrust line* should be kept within the 'middle third' of a wall because, if cracks appeared, the wall might fall down. This is a sound conservative principle which makes for safety and ought to be observed, but, in this permissive age, I am afraid that it seldom is. Anyone who looks at a modern housing estate or a new university cannot help seeing that the walls are full of cracks, and, where there is a crack, there must once have been a tension stress. However, although these cracks do a good deal of damage to the plaster-work and interior decoration,† they seldom constitute any danger to the stability of the main structure.

The basic condition for the safety of masonry is that the thrust line should always be kept well inside the surface of a wall or column.

Dams

Like walls, masonry dams usually fail not from lack of strength but from lack of stability; again they are liable to tip up. The

*There are really several thrust lines, and *all* of them need to be kept inside the surface of the wall.

The passive thrust line. This is the thrust line which results from the weight of the wall itself and of all the things which are permanently attached to it, such as floors and roofs.

The active thrust lines. These are the thrust lines which result, not only from the permanent parts of the building, but also from all the transient loads which might be applied to it by wind pressure or the weight of things like water, coal, snow, machinery, vehicles, people and so on. The shapes of the various active thrust lines define the ways in which a masonry structure can safely be loaded.

†This is one of the reasons for the modern fashion of not plastering the insides of buildings.

sideways thrust on a dam due to the pressure of the impounded water is generally comparable to the weight of the masonry used in its construction. For this reason there are apt to be very large variations in the position of the active thrust line between the 'full' and the 'empty' conditions. With dams, unlike ordinary buildings, one cannot take any liberties at all with the 'middle third' rule. It is quite essential that there should be no cracks of any kind in the masonry, especially on the upstream side. If there are cracks, water under pressure is likely to get inside the structure of the dam and to have two effects, both bad.

The first effect is that the flow of water will damage the masonry; to counteract any seepage, arrangements are generally made to drain the interior of large dams. The second effect is more dramatic. It is that the water pressure within the crack will exert a vertical lifting force (about 5 tons per square foot at a depth of 100 feet, i.e. 0.5 MN/m^2 at a depth of 30 metres) which, acting upon an already rather critical situation, will overturn the dam.

It is probable that the destruction of the Mohne and Eder dams by the R.A.F. in 1943 was accomplished in two stages, separated by a short space of time. In the first stage Barnes Wallis's bombs were dropped against the upstream faces of the dams, where they sank before exploding. When they did explode, the structure of the dams would have been cracked deep down, and after a short delay the actual overturning of the dams themselves was caused by the penetration of high-pressure water into the cracks. Those who have read accounts of the operation will remember that there was an appreciable pause between the explosion of the bombs and the visible failure of the dams. The breaching of these dams, of course, did an immense amount of damage in the Ruhr.

The failure of a dam in peace-time is an engineer's nightmare. Even if the dam is made, not from stone, but from unreinforced concrete, it will be unwise to count upon any reliable tensile strength. Thus in all unreinforced dams the thrust line must not move upstream beyond the 'middle third' when the dam is empty nor downstream of it when it is full, and it is just as well to leave something in hand. These requirements usually result in the tapering, asymmetrical shape with which most of us are familiar (Figure 11).

However, dams are expensive in relation to the value of the water which they impound, and engineers are continually looking for cheaper ways of making them. A considerable saving in the weight and cost of cement can generally be achieved by reinforcing the concrete with steel rods, especially if the reinforcement is under tension. However, unless the reinforcing rods are anchored to solid rock beneath the foundations of the dam, there is a real

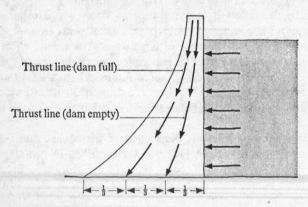

Thrust line (dam full)

Thrust line (dam empty)

$\leftarrow \frac{1}{3} \rightarrow \leftarrow \frac{1}{3} \rightarrow \leftarrow \frac{1}{3} \rightarrow$

Figure 11.   Unreinforced masonry dam.

danger that the whole dam, reinforcements and all, will be uprooted and overturned.

One way of dealing with the situation is shown in Figure 12. Here simple vertical steel tie-rods are anchored to the rock beneath and carried up through the concrete to the top of the dam, where they are tensioned by means of a jacking arrangement. It will be seen that these rods are really doing the same job as the angels and pinnacles in a cathedral. Of course, all traditional heavy masonry may be regarded as a structure which is 'pre-stressed' by its own weight. No doubt a line of heavy statues along the top of a dam would be effective and might look rather nice, but I am afraid they would turn out to be more expensive than the steel rods.

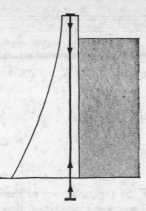

Figure 12. Reinforced dam. A thinner, cheaper dam can sometimes be achieved by using pre-tensioned steel rods anchored in the rock beneath. This is equivalent to extra weight on top of the dam and so restricts the movement of the thrust line.

Arches

Although the arch is not quite as old as masonry itself, it is certainly very old. There is evidence of fully developed brick arches, both in Egypt and in Mesopotamia, going back to about 3,600 B.C. The stone arch seems to have evolved separately, and possibly independently, from the idea of 'corbelling', that is to say, building out the masonry step-wise from each side until the stones met in the middle. The vaulted chambers (Plate 5) deep under the walls of the Mycenaean city of Tiryns – which were old when Homer marvelled at them – are roofed in this way. The postern gate in these immense walls (Plate 6) might be regarded as a development of corbelling. It was probably built before 1,800 B.C.

However, the corbelled* or the semi-corbelled arch, like the gate at Tiryns, is rather a crude affair. Arches soon developed a construction in which the bricks or stones of the arch-ring are made slightly wedge-shaped and are called 'voussoirs'. The various parts of a traditional arch are shown in Figure 13.

*The true arch seems to be an old-world development. The indigenous civilizations of Mexico and Peru built their large buildings using only the corbelled arch.

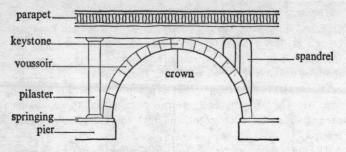

Figure 13. Various parts of an arch.

The voussoir at the top or crown of the arch is called the 'keystone' and is sometimes made larger than the rest. Although poets, politicians and other non-technical people have attributed special qualities to real and figurative keystones, in fact the keystone is functionally no different from all the other voussoirs, and its distinction, if it has any, is purely decorative.

The structural function of an arch is to support the downward loads which come upon it by turning them into a lateral thrust which runs round the ring of the arch and pushes the voussoirs against each other. The voussoirs, naturally, push in their turn against the abutments or springings of the arch. The manner in which this process works is pretty clear from common sense (Figure 14).

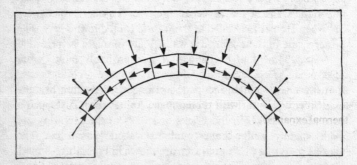

Figure 14. An arch collects the vertical loads and turns them into lateral ones. These lateral loads run round the arch ring and are reacted by the abutments.

The arch ring, with its voussoirs, is very much like a curved wall, and the position of the compressive loads at each joint can be indicated by a thrust line in the same sort of way. In this case the thrust line has to curve round and follow, more or less, the shape of the arch. We shall talk about thrust lines in arches in the next chapter; for the moment let us accept that there *is* a thrust line. Also, as with the wall, we can assume that the voussoirs cannot slide over each other and that the joints cannot take tension.

The joints between the voussoirs will behave in much the same way as the joints between the stones in a wall. If the thrust line strays beyond the 'middle third' then a crack will appear; also, if the thrust line moves to the edge of a joint, that is to say, to the boundary of the arch ring, then a 'hinge' will develop. What makes the arch dramatically different from a mere plebeian wall, however, is that, whereas the wall now falls down, the arch does not. From Figure 15 it can be seen that no fewer than *three*

Figure 15.   An arch can put up with *three* hinge-points without collapsing; in fact many modern arches are deliberately built in this way.

hinge-points can develop in an arch without anything very dramatic happening. In fact a good many modern arch bridges are deliberately built with three hinged joints so as to allow for thermal expansion.

If we really want the bridge to fall down then we shall need *four* hinge-points so that the arch can become in effect a three-linked chain or 'mechanism' which is now at liberty to fold itself up and collapse (Figure 16). Incidentally, this is why, if you want to demolish a bridge – for good or bad reasons – it is best to put the explosive charge somewhere near the 'thirds point' of the arch. This generally involves digging down through the roadway so as to reach the top of the arch ring. Since this takes time, the demolition of bridges behind a retreating army is often ineffectual.

All this means that arches are extraordinarily stable and are not unduly sensitive to the movements of their foundations. If there is any appreciable movement in the foundation a wall will probably collapse*; arches do not much mind, and some sort of distortion is quite common. Clare bridge, for instance, in the

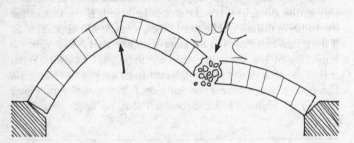

Figure 16. An arch needs to develop *four* hinge-points before it can collapse.

Backs of Cambridge (Plate 7) is very noticeably bent in the middle because of the movement of the abutments. It has been like that for a long time and is quite safe. In the same way arches stand up remarkably well to earthquakes and to other kinds of abuse, such as modern traffic.

Altogether it is not surprising that many of our ancestors were so addicted to arches, for they will probably go on standing up even if you have got all your sums wrong (or not done any sums at all) and, in addition, placed the foundations of the whole thing in a bog – as indeed is the case with several of the English cathedrals.

*This is the rationale of mining or sapping under fortress walls during siege warfare. When the end of the tunnel was beneath the foundations of the wall its roof was supported by wooden props. At an appropriate moment a fire was lit so as to burn through the props, when it was hoped that the wall would collapse. The function of both wet and dry moats was chiefly to prevent sapping.

It is noticeable that, in ruins, it is generally the arches which have survived best. This is partly due to the inherent stability of arches, though it may well have more to do with the fact that the wedge-shaped stones of the voussoirs are less attractive to the local peasantry than the rectangular ones in the walls. The preservation of the round columns of many Greek temples, long after the ashlar of the walls has been stolen, is no doubt due to similar causes.

It is generally easier to keep the thrust line well inside a wall or an arch if the masonry is thick; but of course solid brick and stone-work are expensive. To get extra thickness at a low price the Romans introduced mass concrete. This was usually made by mixing pozzolana (pulvis puteolanis) – a natural earth which is fairly common in Italy – with lime and adding sand and gravel.

If walls and arches are made thicker they are generally more stable and may not need to be made so heavy. If less weight of material has to be transported and handled, then the cost of construction is likely to come down. Vitruvius (fl. 20 B.C.), who was a very distinguished architectural writer as well as an artillery officer, tells us that in his day low-density concretes were often made by incorporating pumice powder. The great dome of the Hagia Sophia in Constantinople (A.D. 528) is made in the same way.

Reduction of weight and cost can be taken still further by incorporating empty containers of one sort or another in the concrete. In the ancient world the very extensive and prosperous wine trade was carried on by means of amphorae. These large earthenware containers were strictly non-returnable and they tended to accumulate in embarrassing quantities. The obvious solution was to cast them into concrete, and in fact many late Roman buildings are made in this way. In particular, the beautiful early Byzantine churches at Ravenna are said to be composed largely of disposable empties.*

*The famous Bristol Channel pilot cutters (c. 1900) were ballasted with concrete which was run into the bilges. The concrete amidships, which needed to be heavy, was made up with scrap-iron and boiler-punchings. The concrete in the ends of the ship, which had to be light, was filled with empty beer bottles. For the plinths of statues and urns in my garden I generally use a mixture of old chicken-wire, empty wine bottles and concrete; it seems to work very well.

Scale, proportion and safety

Although some structures are alleged to be sustained by the Power of Faith and others to be held together entirely by paint or rust, unless a designer is totally irresponsible, he likes to have some kind of objective assurance about the strength and stability of whatever he is proposing to make. If one is unable to do the right sort of modern calculations then the obvious thing to do is either to make a model or else to scale up from some previous smaller version of the structure which has proved to be successful.

This, of course, is pretty well what people used to do down to quite recent times. Perhaps they still do. The difficulty is that models are all very well if one just wants to see what the thing will look like, but they can be dangerously misleading if they are used to predict strength. This is because, as we scale up, the weight of the structure will increase as the cube of the dimensions; that is, if we double the size, the weight will increase eightfold. The cross-sectional areas of the various parts which have to carry this load will, however, increase only as the square of the dimensions, so that, in a structure of twice the size, such parts will have only four times the area. Thus the stress will go up linearly with the dimensions, and, if we double the size, we double the stress and we shall soon be in serious trouble.

The strength of any structure which is liable to fail because the material breaks cannot be predicted from models or by scaling up from previous experience.

This principle, which was discovered by Galileo, is known as the 'square-cube law' and it is one good reason why vehicles and ships and aircraft and machinery need to be designed by proper modern analytical methods. This is probably why such things were so late in developing, at least in their modern forms. However, we can neglect the square-cube law with most masonry buildings because, as we have said, buildings do not normally fail by reason of the material breaking in compression. The stresses in masonry are so low that we can afford to go on scaling them up almost indefinitely. Unlike most other structures, buildings fail because they become unstable and tip up; and for any size of building this can be predicted from a model.

Looking at the problem philosophically, the stability of a building is not different from the stability of a balance or a weighing machine such as a steelyard (Figure 17). The upsetting moments on both sides will be as the fourth power of the dimensions; if we scale up, everything remains in balance. Thus, if a small building stands, a scaled-up version of it can also be relied

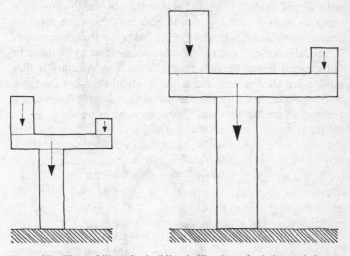

Figure 17.   The stability of a building is like that of a balance; it is not affected by scaling up.

upon to do so, and the 'mystery' of the medieval builders consisted in reducing this experience to a series of rules and proportions. However, that they also used models – sometimes 60 feet (18 metres) long – made from masonry or plaster is well established. This mode of procedure generally worked even for structures of incredible complexity, such as Rheims Cathedral (Figure 18).

The Greeks of the classical period abandoned the arch for most of their serious architecture, preferring to use stone beams or lintels. In such beams the tensile stresses were relatively high and often too near the limit for safety. A considerable number of these architraves were cracked, even in ancient times. This is why iron reinforcement was used in the marble beams of the

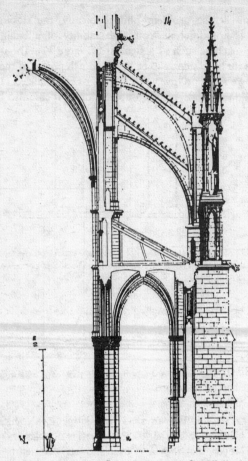

Figure 18.   Rheims Cathedral: flying buttresses (after Viollet-le-Duc).

Propylaea, for instance. What saved the Doric temple from structural collapse was that the stone beams were short and deep and, as they cracked, they turned themselves into arches (Figure 19, Plates 8 and 23).

Greek trabeate* architecture required very large blocks of stone. When civilization decayed, the transportation of large

* From the Latin *trabs*, a beam.

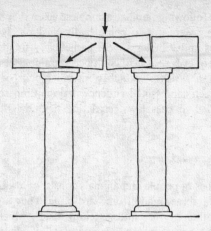

Figure 19.   If a short stone lintel or architrave cracks on the tension face, it may turn itself into a three-hinge arch and continue to support the load.

masses became increasingly difficult, and this may have been one, strictly practical, reason why the medieval builders favoured Gothic arches and vaults, which can be made from quite small stones.

As Sir John Soane pointed out nearly 200 years ago in his lectures on architecture, in spite of the limitations of stone beams, the size of ancient buildings was often greater than that of corresponding modern ones. The Parthenon, for instance, is considerably bigger than St Martin-in-the-Fields. Nevertheless, the Parthenon – about 230 by 100 feet (69 by 30 metres) – is small compared with Hadrian's temple of the Olympian Zeus close by, which measures 359 by 173 feet (108 by 52 metres) and would fill most of Trafalgar Square (Plate 8). Yet Hadrian's temple, in its turn, is dwarfed by the walls of the Acropolis, which tower high above it. Again, for sheer size, many of the Roman bridges and aqueducts are impressive by any standards.

These ancient constructions have more often been destroyed by men than by Nature and some of them are still in good condition today. However, in all these works the ancients were more or less following familiar examples; when they were unable to do so they were apt to come badly unstuck. Not only are ancient ships and vehicles almost pathetically small and fragile to our modern eyes, but new and unconventional buildings such as the Roman *Insulae* – which were tall blocks of flats – fell down with such depressing frequency that the Emperor Augustus was compelled to pass laws restricting their height to 60 feet (18 metres).

On backbones and skeletons

The backbones of people and animals consist of a series of short, drum-like vertebrae, made from hard bone. They are separated from each other by the 'intervertebral discs', which are made from comparatively soft material and thus allow a limited amount of movement between the vertebrae. As a rule, the spine is subject to an overall compression arising both from the weights it has to carry and also from the pull of the various muscles and tendons.

In young people the material of the discs is flexible and tough, and it can withstand considerable tensile stresses if it has to. So much so that, if the spine is damaged by tensile forces, fracture is likely to occur in the bone rather than in the discs. After the age of about twenty, however, the disc material gets progressively less flexible and also considerably weaker in tension. As we get more venerable, therefore, we approach a situation in which our backbone is getting rather like a column in a church or a temple. The vertebrae represent the stone drums and the discs the weak mortar. Although the discs can still, at a pinch, take a certain amount of tension, this is, on the whole, a situation to be avoided.

Therefore, for middle-aged people, it is wise to keep the thrust line as near the middle of the backbone as possible. This is why there is a right and a wrong way of lifting a heavy weight. If we lift the weight in the wrong way, excessive tensile forces are set up in the joints and one of these may break. The result is likely to be a 'slipped disc' or one of the other manifold and rather mysterious back troubles which we include under the name of 'lumbago' – which is apt to be surprisingly painful.

In so far as a backbone behaves like a wall or a masonry column and departure from the 'middle third rule' represents some sort of limiting condition, then the same kind of rules apply to scaling up an animal as we have seen apply to scaling up a building. Thus if we start with a small animal and progressively increase its size, the necessary thickness of the vertebrae will remain in due proportion. Most of the other bones, however, such as the ribs and the bones of the limbs, are subjected chiefly to bending – rather like the lintels of a temple – and the loads upon them are likely to be proportionate to the mass of the animal. It follows, therefore, that such bones have to be made disproportionately thicker.

If we look in a museum at the skeletons of a series of similar animals of increasing size, such as monkeys, it does appear that, whereas the dimensions of the vertebrae of little monkeys and middle-sized monkeys and gorillas and men are roughly in proportion to the height of the animal, the limb bones and, especially, the ribs become very much thicker and heavier, for the size of the animal, as the scale increases (Plate 9).

In this respect Nature seems to be cleverer than the Roman architects, who, as they increased the size of their temples, abandoned the rather stocky Doric proportions and built, as a rule, in the florid Imperial Corinthian style, with slender architraves which frequently broke.

Chapter 10 Something about bridges

– or Saint Bénezèt and Saint Isambard

> *London Bridge is falling down,*
> *Falling down, falling down;*
> *London Bridge is falling down;*
> *My fair lady.*
>
> *Build it up with brick and stone,*
> *Brick and stone, brick and stone;*
> *Build it up with brick and stone;*
> *My fair lady.*
>
> *Set a man to watch all night,*
> *Watch all night, watch all night;*
> *Set a man to watch all night;*
> *My fair lady.*

The more we think about this familiar nursery rhyme, the more eerie it appears to be. Though it cannot be traced with certainty much before the seventeenth century, it is undoubtedly very much older, and the *Oxford Dictionary of Nursery Rhymes* devotes several rather gruesome pages to it. All over the world bridge-building used to be associated with children's dances – *on y danse, on y danse, sur le pont d'Avignon* – and with human sacrifices which are not just legends. At least one child's skeleton has been discovered immured in the foundations of a bridge.* Perhaps for this reason, special orders of bridge-building friars – Fratres Pontifices – were founded during the Middle Ages in various parts of Europe. They produced a saint, St Bénezèt, who

---

*At the Roman fort at Lowbury Hill, in Berkshire, a mile or so from where I am writing this chapter, the body of a woman was discovered concreted into the foundations. The practice has lasted into modern times. In 1865 it was alleged that in Ragusa Christian children were being kidnapped by Mohammedans in order to immure them in the foundations of fortifications. Even in England, as late as 1871, a certain Lord Leigh was seriously suspected of having built an 'obnoxious person' into the foundations of a bridge at Stoneleigh in Warwickshire.

is supposed to have designed the pont d'Avignon. Like Telford, in a later age, he had been a shepherd boy, and it is rather a nice thought that, dispensing with the sacrifices, he kept the children's dances and the tune to which French children dance to this day. The French branch of the order of bridge-building friars had a monastery, near Paris, with the charming name of Saint Jacques-de-Haut-Pas.

In practical terms, the purpose of a bridge is to enable heavy objects, such as vehicles, to cross over some kind of gap or chasm. Provided that the weight is supported in a safe manner it usually does not matter very much by what technical means this is done. As it turns out, there is a very considerable variety of structural principles which can be employed.

The method which is actually chosen in any given case depends not only upon the physical and economic conditions but also upon the fashion of the day and the whim of the engineer. Almost every conceivable way in which a bridge could possibly be made has actually been tried, at one time or another, for making real bridges. One might have supposed that one approach to the problem would have turned out to be the 'best' one and would have come to be generally accepted, but this is not the case; and the number of structural systems which are in common use seems to increase as time goes on.

In civilized countries bridges are littered about the landscape in generous numbers and in a rich variety; they provide a very interesting display of different structural principles. With most other artefacts the vital structure is hidden away behind panelling or insulation or wiring or gadgets of one kind or another and is not easily seen or inferred. One virtue of bridges is that both the structure and the way in which it works are clear for all to see.

Arch bridges

Arch bridges have always been popular and, in various forms, they are still very much in fashion. A simple masonry arch can quite safely be built with a span of well over 200 feet (60 metres). For most sites, if there are objections they are likely to be associated with the cost or with the rise of the arch and with the load on the abutments or foundations.

If we are concerned with the plain, semi-circular masonry arch which was widely used in Roman and medieval times, then one of the facts of life is that the rise of the arch must be about half its span. Thus a 100 foot span will call for a rise of at least 50 feet – in practice rather more. This is all very well if the bridge is spanning a ravine which is more than 50 feet deep, because the arch can then be sunk so that its crown is level with the roadway on either side. However, if the bridge is to be built on flat ground, then we have the alternatives, on the one hand, of a hump-backed bridge, which is inconvenient and dangerous, or, on the other, of having to build long and expensive sloping approaches.

The problem became particularly important with the coming of railways, because trains don't like hump-backed bridges – or indeed gradients of any kind – and the expense of earth-moving to make embankments for a flat approach is a serious matter. One way of getting round the difficulty, at least to a certain extent, is to build a rather flat arch with considerably less rise. In 1837, faced with the problem of getting the Great Western Railway across the Thames at Maidenhead, Isambard Kingdom Brunel built a bridge of two brick arches, each with a span of 128 feet and a rise of only 24 feet (Plate 10).

Both the public and the experts were horrified, and the papers were full of letters prophesying that the bridge would never stand. To keep the correspondence and the publicity going, and perhaps to gratify his sense of humour, Brunel delayed removing the wooden centering or false-work on which the arches had been erected. Naturally, it was said that he was frightened to do so. When, after about a year, the centering was destroyed in a storm, the arches stood perfectly well. Brunel then revealed that the centering had, in fact, been eased to a clearance of a few inches soon after the brickwork was in place and had been doing nothing at all for many months. The bridge is still there today, carrying trains about ten times as heavy as Brunel ever intended.

When we flatten the shape of an arch, so as to reduce the rise in proportion to the span, the compressive thrust between the voussoirs of the arch-ring is considerably increased, as we should expect. However, the compressive stresses are still, as a rule, well below the crushing strength of the masonry, and the voussoirs of the arch are seldom in any danger of being broken, although the deflections which occur when the arch settles after the centering is removed may be quite large and often amount to several inches.

Any real damage to a 'flat' arch, however, is most likely to be a consequence of the greater thrust which must come upon the abutments. If the foundations are of a solid material, such as rock, all will be well, but if they are upon soft ground there may be serious trouble if they yield too much. Unfortunately, the need for a long, flat arch is most likely to occur when we are bridging rivers which flow across a level and boggy country.

It is for these reasons that bridges are often built with many small arches; in fact nearly all long medieval bridges are multi-arch bridges. The objection to this way of doing things is that the cost of building the supporting piers – usually under water and often on soft ground – is high, and, furthermore, the numerous piers and narrow arches obstruct the channel and may cause floods and danger to navigation.

Cast-iron bridges

Some of the objections to arched bridges can be got over by making them from less traditional materials. By the 1770s people like John Wilkinson (1728–1808) – who had greatly cheapened the manufacture of cast iron by improvements to blast-furnaces – began to cast voussoirs from iron. Cast iron is a totally different kind of material from wrought iron and steel because, unlike these substances, it is very brittle. It resembles stone in being strong in compression but weak and unreliable in tension, and so, in building construction, it has to be treated rather like masonry.

An advantage of cast iron is that it is possible to cast architectural members, such as voussoirs, in the form of an open, trellis-like framework, so that there can be an enormous reduction in weight, as compared with traditional masonry. Furthermore, it is generally cheaper to cast iron than to carve stone, and, before taste degenerated around the time of the first Reform Bill, these iron castings were often of very attractive shape.

The benefit of cast iron to bridge-building was two fold. In the first place, there was a saving in labour and transport costs; but more significantly the reduction in the weight of arches diminished the magnitude of the thrust upon the abutments and thus enabled engineers to build flatter arches with cheaper foundations.

Curiously, one of the first people to take advantage of this technique was the American Thomas Paine (1737–1809), who is famous in the history books as the author of *The Rights of Man*. Paine planned to build a great cast-iron bridge, which he had designed himself, across the Schuylkill river, near Philadelphia. He came to England to order the castings, and while they were being made he decided, as a supporter of the French Revolution, to pay a visit to his Jacobin friends in Paris. These gentlemen put him in prison and very nearly guillotined him. He was just saved by the fall of Robespierre.

As a result of the delay, Paine's finances collapsed and the castings were sold off to build a bridge over the Wear at Sunderland. The arch, which was finished in 1796, had a clear span of 236 feet with a rise of only 34 feet. The reason why Brunel did not use cast iron for the Maidenhead bridge forty years later was probably that he was afraid that the vibrations of the trains would crack the brittle cast iron. In any case, his brick arches worked very well.

During the nineteenth century a great many cast-iron arch bridges were built. Although nearly all of them were successful, the method is scarcely ever used today, chiefly because there are now cheaper ways of doing the same job. Unfortunately, a very flat cast-iron arch looks, superficially, rather like a beam (see Chapter 11). Structurally the two are quite different, since the arch is, or should be, entirely in compression, whereas the underside of a beam is in tension. *If* the material can be relied upon to carry tensile stresses, then a beam is often lighter and cheaper than an arch, for a comparable service.

Some of the early engineers, notably Robert Stephenson (1803–59), were tempted by this prospect of economy to venture into using cast-iron beams. Because of Robert Stephenson's out-standing professional reputation the railway companies were persuaded to build several hundred cast-iron beam bridges. However, as we have said, cast-iron is weak and treacherous in tension, and these bridges turned out to be very dangerous indeed. In the end, every one of them had to be replaced and the expense to the companies was naturally severe.

The arch bridge with suspended roadway

A modern tendency in building large arch bridges is to use a suspended roadway. If we split the arch-ring into two parallel elements, which are made from steel or reinforced concrete, then we can hang the roadway from the arches, at any level we like, in much the same way as is done in suspension bridges (Figure 1). There is then, of course, no restriction on the rise of the arch.

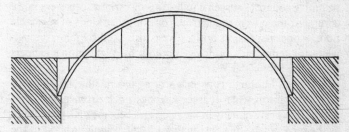

Figure 1. Arch with suspended roadway.

The Hell Gate bridge in New York (1915), which is 1,000 feet (300 metres) span, and the Sydney Harbour bridge (1930), which has a span of 1,650 feet or 500 metres, are steel bridges of this type. In such bridges the main loads are carried entirely in compression in the arches, and the hanging roadway is free from longitudinal stresses. In big bridges the thrust upon the abutments is therefore considerable, and very reliable foundations are needed. Both the Hell Gate and the Sydney Harbour bridges are founded on solid rock.

Suspension bridges

Masonry arches have a number of advantages. As we have seen in the last chapter, they are comparatively easy to design, since one can generally scale up from previous experience quite safely. In fact, as Professor Heyman remarks, it is very difficult to design an arch which will actually fall down. This feat was, in fact, achieved by a certain William Edwards at Pontypridd in 1751, but I do not think there is any record of its happening since. Again, arches are not unduly sensitive to a reasonable amount of movement in the foundations. However, foundations of some sort there must be; and on soft ground they have a way of being both troublesome and expensive.

Furthermore, although the maintenance cost of masonry is usually low, the first cost has always been high, and this is particularly the case with large bridges, which need elaborate centering during erection. For these reasons there has always been a demand for something cheap and cheerful in the bridge line. In primitive countries suspension bridges of various sorts were fairly common; these were made from rope or other kinds of vegetable fibre. Rope suspension bridges were also used by military engineers for temporary bridging, notably by Wellington's sappers during the Peninsular War.

However, although rope is a strong and reliable material for carrying tension when it is new, ropes made from plant fibres deteriorate fairly quickly in the open and become undependable – as the more interesting personalities in the neighbourhood of the bridge of San Luis Rey discovered.* For a permanent suspension bridge, cables of iron or steel are necessary. Cast iron was far too brittle and steel was not commercially available until relatively recently, but wrought iron is fairly strong and very tough; also it is exceptionally resistant to corrosion.

Although a footbridge 70 feet (20 metres) long, made with iron chains, was erected over the Tees in 1741, wrought iron was generally too expensive to be used at all widely in bridge-building until the puddling process† was introduced about 1790. After this,

* *The Bridge of San Luis Rey*, Thornton Wilder (1927).
† *The New Science of Strong Materials*, Chapter 10.

wrought-iron chains became comparatively cheap. In the Tees bridge the flooring was attached directly to the chains in the primitive manner, so that the bridge was impassable to vehicles and must have been both steep and alarming for pedestrians. The modern system of supporting the cables from high towers and hanging the roadway below the cables (Figure 2) was invented by

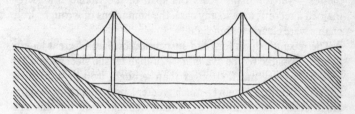

Figure 2. The modern form of suspension bridge, with a level roadway hung from the cables, was invented by James Finlay about 1796.

James Finlay, of Pennsylvania, who began to build bridges of this kind around 1796.

The combination of a suspended, level roadway with the availability of wrought-iron chains at a reasonable price made the suspension bridge an attractive proposition for carrying wheeled traffic over wide rivers. For many situations these bridges were much cheaper and more practical than large masonry bridges. The idea was taken up very actively in many countries, and especially by Thomas Telford, whose bridge across the Menai Straits (Plate 11) was finished in 1825; it has a centre span of 550 feet (166 metres), by far the longest then in existence.

Telford's chains, like all the suspension chains used in bridges at that time, were made from flat plates or links, joined by bolts or pins, very much like the links of a modern bicycle chain. The concentration of stress at the pin joints calls for a tough and ductile material, such as wrought iron, and indeed chains of this type have been very successful and have seldom given any trouble. Although wrought iron is reliable in tension it is not especially strong, and Telford wisely kept the highest nominal stress in his chains down to about 8,000 p.s.i. (55 MN/m²), which is less than a third of the breaking stress.

In these circumstances a great deal of the strength of the chains was devoted to supporting their own weight, and Telford was of the opinion that the Menai bridge represented about the maximum safe span for a suspension bridge, using the materials of the day. Although Brunel eventually showed that Telford was being rather cautious – Brunel's Clifton bridge has a span of 630 feet or 190 metres – yet for many years the span of the Menai bridge remained a record; and, in any case, the limitations of wrought-iron chains were clearly within sight.

The recent fashion for road suspension bridges of great length is made possible by the availability of high tensile steel wire. This material is very much stronger than wrought iron or mild steel and can therefore support a much greater length of its own weight. High tensile steel is more brittle than wrought iron, but this can be accepted, since the cable is continuous and does not have to have links with pinned joints, which are particularly vulnerable to cracking. Again, instead of having only three or four plate links in parallel with each other in each element of a chain cable, the wire cables are woven from many hundred separate wires, so that the failure of any individual wire is not likely to be dangerous (Plate 12).

As an example of the sort of thing one can do nowadays, the new Humber motorway bridge has a clear span of 4,626 feet (1,388 metres), which is over eight times the length that Telford thought practicable. This is made possible by the fact that the suspension wires operate, quite safely, at a working stress of 85,000 p.s.i. or 580 MN/m^2, which is more than ten times the stress in Telford's wrought-iron chains.

Thrust lines in arches and suspension bridges

The cables of a suspension bridge take up the best shape automatically, because a flexible rope has no choice but to comply with the resultant of all the loads which are pulling on it. We can therefore determine the shape of the cables for a suspension bridge either by loading a model of it, as Telford did, or else by means of a fairly simple exercise with a thing called the 'funicular polygon' on the drawing board. This is useful in designing suspension bridges – for instance we need to know the right lengths for the hangers for the roadway – but it is also useful in designing arches.

If we look at a suspension bridge and then at an arch, it does not need much imagination to see that the suspension bridge is really an arch turned upside down – or vice versa. In other words, if we change the sign of all the stresses in an arch, that is, if we turn all the compressions into tensions, then these tensions could be carried by a single curved rope, which may be regarded as defining a 'thrust line' in tension. By doing this we can arrive, comparatively painlessly, at the compressive thrust line for an arched bridge or a vaulted roof.

When we do so we may get various shapes of thrust line which will vary a bit according to the details of the loading, for instance the presence or absence of traffic on the bridge. Any of these thrust lines will be safe, provided that it lies wholly within the intended shape of the arch-ring; if not, not. It is sometimes said, by slightly superior people, that the thrust line of an arch obtained in this way has the shape of a catenary, and that a round arch is therefore 'wrong'. This is by no means always the case, and in many instances the thrust line is quite near enough to an arc of a circle to justify the Romans in their highly durable semi-circular arches. However, if one wants to make a really thin arch – as is the custom with modern reinforced concrete bridges – then one had better get the shape just right, for there is very little room for the thrust line to wander about.

The development of the bowstring girder

Although the suspension bridge got off to a flying start at the beginning of the nineteenth century, its development was interrupted for about a hundred years by the coming of the railways. Most of the 25,000 major bridges which were built in England during the Victorian era were railway bridges. The suspension bridge is a highly flexible structure, and it is liable to deform dangerously under large concentrated loads. This characteristic does not matter very much for road bridges,* but trains are

* All Telford's bridges were road or canal bridges. The Americans made fairly extensive use of suspension bridges for canal aqueducts; the water

generally about a hundred times as heavy as carts or lorries, and so the deflections they cause are likely to be a hundred times as great and therefore unacceptable. The few railway suspension bridges which were built in England were conspicuous failures. The Americans, who had wider rivers, and at that time less money and more faith, persisted with them for a while but had to give most of them up in the end.

There was therefore a need for bridges which were not only light and cheap but also rigid and suitable for large spans. This led to the development of what is called the 'tied arch' or 'bowstring girder' (Figure 3). An arch, of course, is pretty rigid, but it

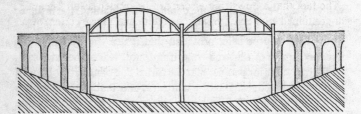

Figure 3.   The bowstring girder, or tied arch, relieves the abutments of lateral thrust. It was popular with Victorian railway engineers.

thrusts outwards on its abutments with a very considerable force. This may not matter if these abutments consist of nice firm rock, but it is awkward in many of the situations which may arise in railway construction. It is particularly inconvenient if it is required to perch an arch, or a series of arches, on top of tall and slender piers which may be in no position to resist large lateral loads.

However, this is just what the Victorian engineers so often wanted to do, for they frequently carried their railways boldly across deep valleys, sometimes at a height of 100 feet or more. One way of solving the problem is to tie the two ends of the arch together by means of a tension member. This can be done by

channel was carried in a suspended wooden flume. Naturally there was no change of net load – and therefore no change of deflection – when a barge passed over the bridge.

using a suspended roadway, which in this case is made to work for its living: the roadway itself is put into tension.

The bowstring girder looks superficially like an ordinary arch with a suspended roadway, but its manner of working is quite different. Now there is no sideways push or pull upon the foundations, which have only to support the vertical downward load arising from the actual weight of the girder and any vehicles which may be on it. In fact the whole affair can be mounted on rollers instead of on rigid foundations, and this is often done, mainly to allow for thermal expansions and contractions in the metal. Since such girders produce no lengthwise thrust, they can be mounted on top of relatively narrow masonry columns.

The fact that a bowstring girder can be treated as an integral, self-contained unit may greatly facilitate the construction of a large bridge, because it is possible to assemble the girders at ground level, on some site away from the bridge itself. They can then be floated out to the piers on rafts and raised into position by means of jacks. This is just what Brunel did with the spans of the Saltash bridge. As we shall see in the next chapter, the tied arch is really yet another member of the prolific family of 'trusses' or lattice girders with which structural engineering is so thickly populated.

Chapter 11   The advantage of being a beam

– with observations on roofs, trusses and masts

> *Solomon . . . built the House of the Forest of Lebanon,*
> *a hundred cubits long, fifty broad, and thirty high,*
> *constructed of four rows of cedar columns, over*
> *which were laid lengths of cedar. It had a cedar roof,*
> *extending over the beams, which rested on the*
> *columns, fifteen in each row; and the number of the*
> *beams was forty-five.*

1 Kings 7.1–3 (New English Bible)

A solid roof over one's head is one of the prime requirements of a
civilized existence, but permanent roofs are heavy and the problem
of supporting them is really as old as civilization itself. When one
looks at a famous and beautiful building – or indeed at any building – it is illuminating to bear in mind that the way in which the
architect has chosen to solve his roofing problem has affected not
only the appearance of the roof itself, but also the design of the
walls and the windows and indeed the whole character of the
building.

In fact, the problem of supporting a roof is essentially similar
in its nature to the problem of making a bridge, with the difference
that, since the walls of buildings are likely to be thinner and
weaker than the piers of bridges, any sideways thrust which the
roof may impose must be considered even more carefully. As we
saw in Chapter 9, if the roof pushes outwards too hard upon the
tops of the walls on which it rests, the line of the thrust in the
masonry will be displaced to a dangerous extent and the walls will
collapse.

Many Roman buildings and practically all Byzantine formal
architecture made use of vaulted or domed roofs. These arch-like
structures thrust vigorously outwards upon their supports, and
in most cases this was catered for by resting the roofs upon very
thick walls within which the thrust line had generally plenty of
room to wander about in safety. As we have seen, these thick

walls were often made from mass concrete, sometimes lightened
and thickened by the incorporation of empty wine jars. Such walls
were structurally stable, and they had the additional advantage of
providing excellent heat insulation in hot climates: a Byzantine
church is often the only cool place in a Greek village. However, it
is not easy to make windows in very thick walls, and such windows as existed in Roman and Byzantine buildings were usually
small and placed high up.

The medieval castles were built pretty much in the Roman
tradition, frequently, as at Corfe Castle, from mass concrete many
yards thick. Such walls were well able to resist the thrusts set up
by the vaulted roofs; and, for military reasons, the defenders did
not really want windows anyway. The earlier Norman or
'Romanesque' churches were not very different, and their thick
walls, little, rounded arches and small windows derive directly
from late Roman prototypes. Most of the early Romanesque
churches were satisfactory enough, and many of them survive
today.* The difficulties and complications arose later on and are
to a great extent associated with the growing fashion for bigger
and better windows.

Understandably, people living in sun-drenched countries do not
feel quite the same about windows as northerners, and even today
many of them seem to dwell, apparently from choice, in a perpe-
tual shuttered twilight. No doubt this is all part of a long Mediterranean tradition, for in Greek and Roman and Byzantine times
such windows as existed were generally small and rather ineffectual.† As far as one can see, this was by no means entirely due
to a shortage of glass.

In northern Europe, even warlike knights and barons did not
want to spend all their time in gloomy and nearly windowless
castles. What they wanted was light and sunshine, and so they
tired of architectural forms based upon dark Roman models. The

---

*Of course, a great many small Norman churches have simple wooden
roofs, but the design of these roofs is often such that they thrust outwards
upon the walls nearly as badly as a stone vault.

†In Pompeii, where the windows are inadequate and the artificial light
must have been bad, the walls of nearly all the rooms are painted either dark
red or black. One wonders why.

cult of windows became an obsession, and, as time went on, builders competed in constructing both halls and churches with larger and larger and ever more splendid windows. The medieval craftsmen may have been hopelessly unscientific but they were sometimes much more creative than we generally recognize. In particular we owe them a great debt for showing us what beautiful and exciting things can be done with windows.

However, much of the effect of an impressive and expensive window is lost if it has to be inserted into a tunnel-like opening in a thick wall. Inevitably, attempts to provide bigger windows set in thinner walls ran into trouble with thrust lines. Norman architec-

Figure 1. King's College Chapel, Cambridge.

ture was basically Roman architecture and cannot be made to do this sort of thing, because it depends for its stability and safety on the use of thick walls. But this did not stop builders from trying, and it has been said of late Romanesque architecture that the question to ask of any particular building is 'not whether, but when, the Great Tower fell'.

Just how clearly the medieval masons appreciated what was

happening is not certain. Most probably their understanding of the situation was muddled and subjective; otherwise they would not have gone on making the same mistakes for several generations. Sooner or later, however, somebody realized that the way

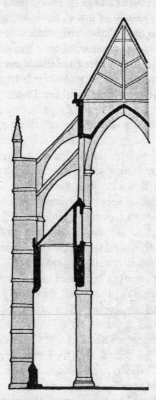

Figure 2. The introduction of side aisles and a clerestory required the invention of the flying buttress.

to deal with a demand for large windows and thin walls was to make use of buttresses, which could prop the wall against the outward thrust of the roof by pushing against it from outside.*

*‘I am not a Pillar, but a Buttress, of the Established Church, since I support it from without’ (Lord Melbourne).

Effectively, buttresses make the wall thicker, and so they do the same job as the Roman wine bottles, only in a different way.

The ordinary solid buttress is really no more than a local thickening of the wall between the windows. Where there is only a single aisle, as in King's College Chapel (Figure 1 and Plate 13), it is very effective. Difficulties arose, however, with side aisles. In order to prop the roof of the nave without unduly shading the clerestory windows, the Gothic masons had to invent the flying buttress (Figure 2). In this case the vertical part of the buttress is separated from the wall by a series of arches, which transmit the thrust without intercepting much of the light.

The decorative possibilities of flying buttresses in conjunction with large windows are very great, and, as we have said, they are still further enhanced by the judicious introduction of statues and pinnacles, whose weight, as the masons must somehow have realized, helps the buttresses in the tricky task of guiding the thrust lines safely down through the lace-like forest of masonry. In the end the windows became so large that not very much actual solid wall was left to support the building. Like a modern mast, these narrow strips of stonework depended entirely upon lateral support. As a tall thin mast relies upon a network of sophisticated rigging, so these slender walls depend entirely for their stability upon the bracing afforded by arches and buttresses.

By whatever mental process all this was accomplished, the structural and artistic achievement was immense. By the time the master masons had created the Gothic buildings of the high Middle Ages, architecture had lost any visible connection with its classical origins. Few things could look much more different than, say, Canterbury Cathedral and a Roman basilica. Yet the line of descent is clear and simple.

Although buildings like these are often very beautiful, they are always horribly expensive, and in any case arched or domed roofs are usually unsuitable for private houses. Rather than using arches it is much cheaper and simpler to support the roof of a building by using beams of one sort or another. If the spaces to be covered are spanned with long poles or joists, then such beams can transmit the weight of the roof from their ends, vertically downwards into the masonry of the walls, without any need to push sideways and outwards. Thus no unwelcome disturbance is caused to the thrust line and so the walls can be made quite thin and will not need buttressing (Figure 3).

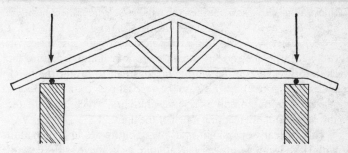

Figure 3. Simply supported roof-truss. This one is shown mounted on rollers to emphasize that there need be no outward thrust upon the supporting walls.

For this reason alone, the beam is one of the most important devices in the whole of structural engineering. In fact, however, the applications of the beam – and of its equivalent, the truss – extend far beyond the roofing of buildings; and beams and beam theory have played a very important part indeed in making technological civilization possible. Similar ideas are also continually cropping up in biology.

The word 'beam' means a tree in Old English, and this usage still survives in tree names like 'whitebeam' and 'hornbeam'. Although nowadays beams are very commonly made from steel or reinforced concrete, for a great many years a 'beam', in the structural sense, implied a baulk of timber, very often a whole tree-trunk. Although it is cheaper and much less trouble to cut down a tree than to build a masonry arch or vault, the supply of suitable large trees is not unlimited and a time arrives when long pieces of timber become scarce. When this happens one may be forced to try to construct roofs from short lengths of material.

Roof trusses

To the modern mind it might seem fairly clear that the most promising way to try to bridge a roof-span using short pieces of timber would be to join the short members together, Meccano-fashion, so as to make a triangulated structure, something like Figure 4. This is really the beginning of a lattice girder. We are all

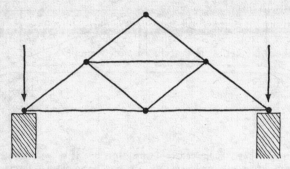

Figure 4.   If long pieces of timber cannot be got, then a roof-truss could be built up, Meccano-fashion, from short pieces.

familiar with lattice girders in steel railway bridges. Any triangulated lattice structure of this kind is called a 'truss'. Like a long solid beam, when a roof-truss is properly designed it allows considerable spans to be roofed economically and without putting any dangerous outward thrusts upon the supporting walls. As with beams and beam theory, the applications of trussing in modern technology are far wider than this and extend to ships and bridges and aeroplanes and to all manner of other structural devices. As we have seen in the last chapter, the tied arch is really another example of the same idea.

However, in the history of architecture the concept of the truss or lattice beam was surprisingly slow in developing. The most primitive form of this idea, the ordinary wooden roof-truss, may seem obvious to us but it took our ancestors a long, long time to think of it. But then they had never seen a railway bridge or played with Meccano. As it turned out, architectural trussing was a late Roman invention, although it never really caught on properly until the Middle Ages. During most of antiquity architects just managed without trusses.

Greek builders never thought of trusses at all. Vastly eminent Athenian architects, such as Mnesicles, who built the Propylaea, and Ictinus, who designed the Parthenon and the Temple of Apollo at Bassae, consciously rejected arches and vaults as a method of roofing their buildings, and yet they failed conspicuously to invent the roof-truss or to devise any really adequate substitute for it. The brilliance of Hellenic architecture seems to come to a stop, rather suddenly, when one gets to the architrave. Greek roofs can only be described as intellectually squalid.

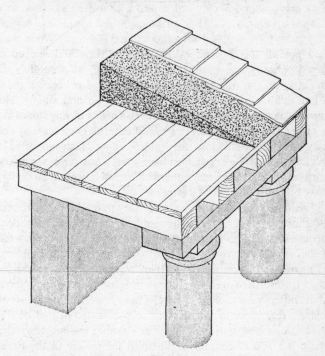

Figure 5.   Roof of an archaic Greek temple.

Simple stone beams or lintels cannot safely be used to span distances of more than about eight feet (2·5 metres); otherwise they are liable to crack. Therefore, in order to provide practicable roofs for temples and other buildings, it was necessary to use wooden beams, in spite of the fact that in classical Greece timber had become nearly as scarce as it is in modern Greece.

In Greek temples for which the necessary number of full-span wooden roof-beams could be found, the beams were simply laid horizontally right across the tops of the walls and of the stone lintels of the peristyle. These beams or joists were then boarded over so as to provide a continuous flat ceiling over the whole area of the building (Figure 5). Naturally this flat ceiling, which was only made from ordinary planks, was anything but weatherproof. So a great mound of clay soil, mixed with water and straw, was heaped up on top of it. For an average-sized temple this pile of clay must have weighed something like 3,000 tons. When they had got all this agricultural material up there and tamped it down properly, the mound was trimmed off as accurately as might be to the triangular shape of a pitched or sloping roof. After this the roofing tiles were brought up and simply laid directly on top of the clay, very much like laying paving stones for a garden path. Presumably one hoped that this great mass of wet clay would dry out before the wooden ceiling which supported it began to rot. When dry, it must have made a wonderful sanctuary for vermin; but the excellent thermal insulation would, no doubt, have been welcome in hot weather.

Frequently, of course, it was necessary to use beams or rafters of shorter length. King Solomon had made special political arrangements* with King Hiram for the supply of cedar from the Lebanon, but even so his roof-beams were only about 25 feet long (7 metres or 17 cubits). Many Greek temple beams were shorter than this. In the Greek temples, as in Solomon's building, these short rafters were supported directly from underneath, by rows of pillars, regardless of architectural convenience. In one of the great Doric temples (c. 550 B.C.) at Paestum in southern Italy, there is a line of columns right down the middle of the nave, dividing it into two equal aisles. This must have made any kind of religious ceremony very awkward. In most of the later temples more seemly and more symmetrical arrangements were generally achieved (Figure 6), but even the interior of the Parthenon was cluttered up with pillars which we should think unnecessary.

The simplest form of roof-truss, which was an 'A'-shaped

* 1 Kings 5 (where there is a strong hint that Solomon had to pay a stiff price).

Figure 6.   The more sophisticated temples of the fifth century managed to support their roofs without the use of trusses.

affair, was developed during the Middle Ages. The horizontal tension member or tie-bar across the bottom of the truss is called by builders the 'collar'. For short spans it was generally easy

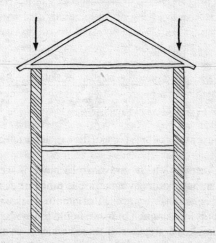

Figure 7.   Simple two-storey house with the collar of the roof-truss level with the tops of the walls.

enough to find timbers for the collar which were sufficiently long to make a simple triangular truss like Figure 7, but for a small two-storied house this arrangement often results in rather clumsy architectural proportions; moreover, a good deal of space may be wasted in the roof. For these reasons builders often attached the collar higher up – in effect putting the upstairs rooms partially within the roof and using dormer windows where necessary. This is all very well, but, if the collar is put high up on the truss, there is a tendency for the rafters to bend or spring outwards under the weight of the roof. This pushes the wall outwards at the same time (Figure 8), very possibly with expensive results. Natu-

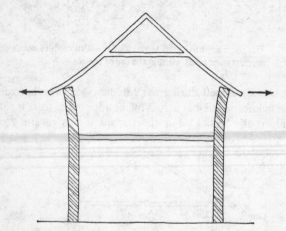

Figure 8. The effect of raising the collar too high in order to save space and cost. (Exaggerated – but not much.)

rally, the higher the collar is placed, the worse the effect is likely to be.

In large medieval halls and churches, which were often of considerable span, roofing was a serious problem. A trussed roof might be cheaper than an arched or vaulted masonry one, but, even if timbers long enough to make full-length tie-bars or collars could be found, the presence of these collars comparatively low down in the building spoilt the architectural effect of the nave or hall, and, in particular, they blocked the view of the great east and west windows. Since people in those days were often so backward as to pay more attention to appearances than to 'efficiency', Continental builders stuck to masonry vaulting, supporting

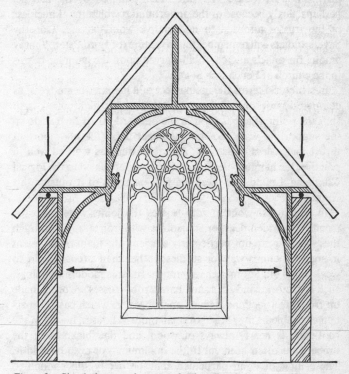

Figure 9.   Simple hammer-beam roof. The effect is to move the point of application of the outward thrust (which results from the distortion of the truss) further down the walls so that it has less effect on the thrust line. At the same time the view of the end window is kept clear.

their arched roofs by means of elaborate and expensive buttressing.

Characteristically, the English builders produced a compromise or palliative type of timber roof, which has been described as 'more ingenious than scientific'. This was the 'hammer-beam' roof (Figure 9). Hammer-beam roofs became comparatively popular for large buildings in England, and they can be seen in Westminster Hall, in many Oxbridge colleges and in some large private houses today. They are much admired by the artistic, perhaps partly because of the opportunities which the 'knuckles' of the trusses afforded to imaginative wood-carvers. Dorothy Sayers addicts will remember the adventures of Lord Peter Wimsey among the angels and cherubim carved upon the hammer-beams in the church of Fenchurch St Paul.*

In structural terms the main effect of a hammer-beam truss, as compared with any similar large truss with a high collar, is to shift the point of application of the outward thrust further down the supporting walls, so that its effect upon the all-important thrust line is less disastrous. Although this has worked well in practice, the hammer-beam truss has never appealed to the logical Continental mind and there are few examples of it outside this country.

In traditional wooden roof-trusses the joints were made by means of wooden pegs, or sometimes with iron straps. Although these joints were not particularly efficient, the main requirement in such structures was for stiffness rather than strength, and so weak joints did not matter very much. In large modern buildings, such as factories and sheds and barns, roof-trusses are often made up from steel sections such as angle-bars, in which case no particular problems may arise. In small modern houses, however, the roof-truss is nearly always of wood, and the thickness of the timbers has often been cut to the minimum – or even beyond it. The ceiling-joists, in particular, may be barely stiff enough to support the ceiling without causing the plaster to crack. If we are tempted to indulge in the fashionable activity of turning a modern attic into an extra bedroom, the most serious problem is likely to be the stiffness of the floor. Although the roof-truss is unlikely to break, the deflections caused by the extra weight of people and furniture may well cause serious and expensive damage to the house. Amateur handymen, please take note.

* *The Nine Tailors* (Gollancz, 1934). But the roof-trusses of the little church of St Swithin at Wickham in Berkshire are decorated with large Victorian papier-mâché elephants.

Trusses in shipbuilding

> *There is a land of sailing ships,*
> *a land beyond the rivers of Cush*
> *which sends its envoys by the Nile,*
> *journeying on the waters in vessels of reed.*

Isaiah 18.1–2 (*c.* 740 B.C.; New English Bible)

As a matter of fact, trusses of various kinds were used and understood by shipwrights for many centuries before builders and shore-going architects got round to the idea. Most histories of ship-building begin with the boats which the Ancient Egyptians made for use on the Nile. As the prophet Isaiah seems to have been well aware, these boats were constructed by tying together several parallel bundles of reeds. Actually, these reed boats, which developed from rafts, date back to long before the time of Isaiah, probably to somewhere between 4,000 and 3,000 B.C. Similar boats are in use today on the White Nile and also on Lake Titicaca, in South America. Since the bundles of reeds naturally tapered towards the ends, a roughly boat-shaped form was achieved more or less automatically. Often the long, wispy, ends of the reed bundles were tied in such a way that they turned upwards so as to provide a vertical decoration at the bow and stern. This feature survives today, sometimes not very much changed in shape, in the high stemposts of Mediterranean rowing boats – especially in the Venetian gondola and the Maltese dghaisa.

Although most of the buoyancy of a ship is provided by the middle part of the hull and comparatively little by the tapering ends, nothing will ever prevent people from putting heavy weights into the ends of a ship. One result of this is that many vessels tend to 'hog' (the two ends tend to droop and the middle of the hull tends to rise). This state of affairs is the opposite of that which exists in roofs and bridges, where the middle of the truss is usually trying to sag below the level of the end-supports. This condition is called 'sagging' by engineers. Although in hogging and sagging the forces and deflections are acting in opposite directions, it is clear that in both cases the beam or truss is being bent and that precisely analogous principles and arguments apply.

Structurally speaking, a ship's hull is a sort of beam, and the effect of the hogging forces on the flexible reed hulls of the Egyptian boats must have been very obvious. A hogged ship is a depressing thing to look at, and this state of affairs needs to be prevented for all sorts of other excellent reasons, so that it was necessary to do something about the situation even in 3,000 B.C. In fact the Egyptians tackled the problem extremely sensibly. They provided their ships with what is now called a 'hogging-truss'. This consisted of a stout rope which was passed over the tops of a series of vertical struts, its two ends being looped under and round the ends of the ship, so as to prevent them from drooping (Figure 10).

Figure 10.   Egyptian sea-going vessel, *c.* 2,500 B.C. This one is built of wood but retains the vertical ornaments at stem and stern characteristic of reed-built boats. The wooden planks are very short and badly fastened, hence this ship also retains the traditional Egyptian hogging-truss. Note the A-shaped mast.

This rope could be tightened by some form of 'Spanish windlass'. The latter device is a skein of cords which can be twisted – and so shortened – by means of a long stick or lever thrust through its middle. Thus the big reed hull could be strained to any degree of

Plate 1

Chapter 2

Each of the four columns which support the 400 foot (120 metre) tower of Salisbury Cathedral is very noticeably bent. Masonry is much more elastic than is generally supposed.

Plate 2

Chapter 4

Stress concentration at a crack tip. The shear stress in a transparent material is revealed by polarized light. The bands in this photograph are, in effect, contours of equal shear stress.

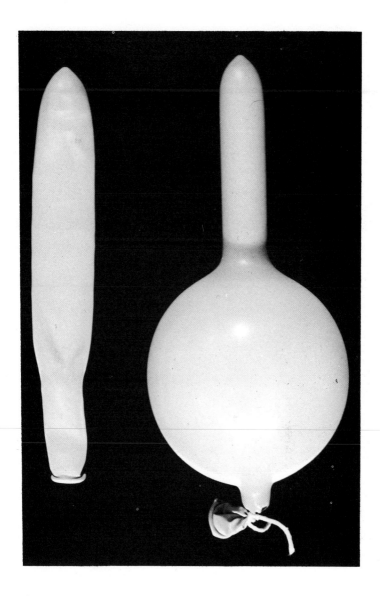

Plate 3

Chapter 8

Rubber has a 'sigmoid' stress-strain curve like Figure 4, Chapter 8. A tube made from such a material will not distend evenly under pressure but will bulge into an 'aneurism'. This is why artery walls do not have rubbery elasticity.

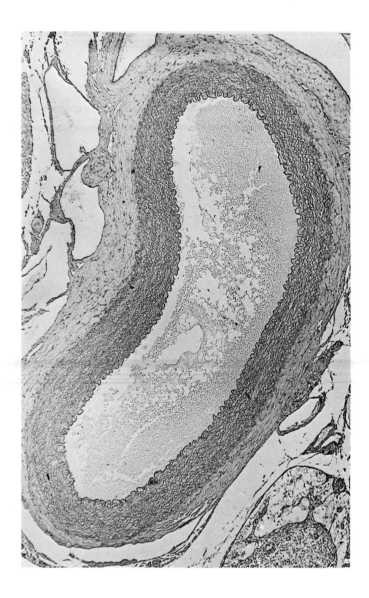

Plate 4

Chapter 8

Artery walls and other living soft tissues have a special kind of elasticity like that in Figure 5, Chapter 8. The artery wall is constructed partly of elastin reinforced by kinked collagen fibres. This helps to produce the required 'safe' type of elasticity. (The artery tends to flatten when it is emptied of blood after death.)

Plate 5

Chapter 9

Corbelled vault at Tiryns (*c.* 1,800 B.C.). Corbelled arches and vaults preceded the true arch.

Plate 6

Chapter 9

Semi-corbelled postern gate at Tiryňs. These walls were old when Homer marvelled at them.

Plate 7

Chapter 9

It is very difficult to get a true arch to fall down. The foundations of Clare bridge, Cambridge, moved a long time ago, but the bridge is perfectly safe though the arch has distorted.

Plate 8

Chapter 9

Part of the enormous temple of the Olympian Zeus at Athens. It was built in the Corinthian style by the Emperor Hadrian about A.D. 138. One of the architraves can be seen to be cracked. Note the walls of the Acropolis, which tower above Hadrian's temple.

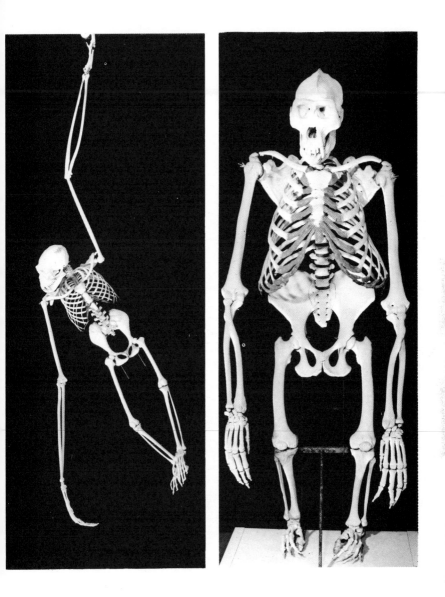

Plate 9

Chapter 9

Skeletons of (a) gibbon and (b) gorilla (to scale). The 'square-cube' law applies more to beams than to columns. Thus as animals get larger, their ribs and limb bones tend to become thicker in proportion to their vertebrae.

Plate 10

Chapter 10

Brunel's Maidenhead bridge (1837) has the longest and flattest brick arches in the world. Many people predicted that the arches would not stand, but they are still there today carrying trains ten times heavier than Brunel's.

Plate 11

Chapter 10

Telford's Menai suspension bridge (1819). The span of 550 feet (166 metres) is approaching the limit for wrought-iron suspension chains.

Plate 12

The Severn suspension bridge. High tensile steel cables with ten times the tensile strength of wrought iron enable bridges nearly ten times as long as Telford's Menai bridge to be constructed.

Plate 13

Chapter 11

Where there are no side aisles, as in King's College Chapel, Cambridge, the buttresses can be carried straight up without further complication.

Plate 14

Chapter 11

H.M.S. *Victory*. Her masts form a superb example of a trussed cantilever structure of very large dimensions.

Plate 15

Chapter 11

The American railways could be built quickly and cheaply because wooden trestle bridges were used very extensively to save the cost of earthworks (*c.* 1875).

Plate 16

Chapters 11 and 13

Stephenson's Britannia railway bridge (1850) used wrought-iron box beams. The trains ran inside the beams. Much trouble was experienced in preventing the thin iron plating from buckling. In front of the bridge are grouped a number of contemporary engineers; Robert Stephenson is seated in the left centre and I. K. Brunel is seated on the extreme right.

Plate 17

Chapter 12

The bias cut, invented by Mlle Vionnet, exploits the low shear modulus and high Poisson's ratio of certain square-weave fabrics in the 45° direction. This is one of the earliest Vionnet bias-cut dresses (A.D. 1926).

Plate 18

Chapter 12

Contemporary square-cut dress (also Vionnet). Note low Poisson's ratio and lack of clinging effect. The vertical creases are caused by the existence of a Wagner tension field.

Plate 19

Chapter 12

Wagner tension field in the fuselage skin of a Fairey
Rotadyne.

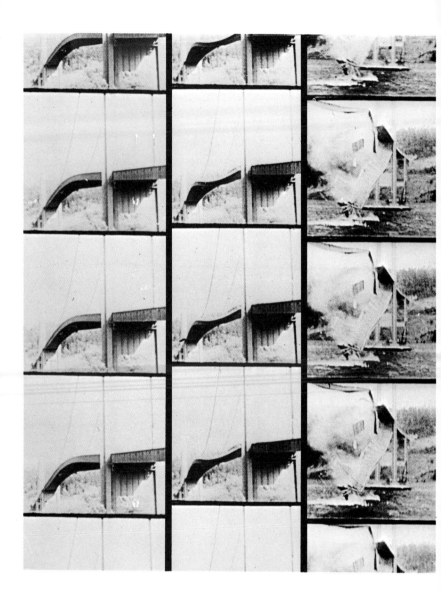

Plate 20

Chapter 15

The Tacoma Narrows bridge is a classic example of a
suspension bridge built with inadequate torsional stiffness.
Known as 'Galloping Gertie' it displayed serious oscillation
when exposed to quite moderate winds, and very soon
wriggled and buckled itself into failure in a wind of only
42 m.p.h.

Plate 21

Chapter 16

The first real mass-production machinery to come into use was the block-making equipment at Portsmouth Dockyard. Both the machinery and the blocks themselves may be regarded as handsome, perhaps as beautiful.

Plate 22

Chapter 16

The classical form of steam yacht developed by George Lennox Watson is one of the most beautiful of all ship conventions. But it is largely non-functional. The ends and especially the bowsprit represent sailing-ship practice. (S.Y. *Nahlin.*)

Plate 23

Chapters 9
and 16

No single photograph can do justice to the Parthenon, but
this picture of part of the south-west corner may give some
slight impression. (Note that the left-hand lintel is cracked;
for this reason the architraves are in triplicate.)

Plate 24

Chapter 16

Unlike the classical Greeks one thousand years later, the Mycenaean Greeks (*c.* 1,500 B.C.) designed their buildings to take account of the low tensile strength of stone. The lintel of the Lion Gate at Mycenae is provided with a triangular block of stone to relieve the tensile loads. The architrave is a single block and carries very little stress.

straightness or vertical curvature which the skipper happened to fancy. As the art of shipbuilding progressed, the Egyptians came to construct their hulls from timber, rather than from bundles of reeds. But, since most of the planks were very short and nearly all of the fastenings might be described as wobbly, the need for the hogging-truss remained.

Greek shipwrights were more advanced than the Egyptian ones and they built the splendid triremes or fighting galleys upon which the sea-power of Athens depended. However, these ships were also built from short lengths of timber, and their light hulls were very flexible and much inclined to leak. For these reasons the Greeks retained the hogging-truss in the sophisticated form which was called the *hupozoma*. This was a substantial rope which ran right round outside the hull, high up and just beneath the gun-wale. Again the *hupozoma* was set up by means of a Spanish windlass which could be adjusted as needed by the helmsman. Since Greek warships fought mainly by ramming each other, they had to be able to withstand a great deal of structural abuse. The *hupozoma* was therefore an essential part of the hulls of these ships; they were unable to fight, or even to go to sea at all, without it. Just as it used to be the practice to disarm modern warships by removing the breech-blocks from the guns, so, in classical times, disarmament commissioners used to disarm triremes by removing the *hupozomata*.

It is quite clear that the Athenian shipbuilders, down in the Piraeus, were familiar with the principles of trussing, and one might well ask why the Athenian architects, such as Mnesicles and Ictinus, did not latch on to the idea for the roofs of their temples. Perhaps the analogy between hogging and sagging never struck them, or perhaps they just never hobnobbed with shipwrights. After all, how many house architects today ever talk to a naval architect?

When the fragile oared fighting galley went out of use, hogging-trusses disappeared. However, the American river steamboats of the nineteenth century were every bit as flexible as the Greek trireme or the Egyptian vessels on the Nile. Their shallow wooden hulls presented exactly the same problems, and the Americans solved these problems in precisely the same way as the ancient

Egyptians did. All the American river steamers were provided with hogging-trusses of the Egyptian pattern. The only difference was that the tension members were made from iron rods, rather than papyrus rope, and they were tightened by means of metal screws instead of a Spanish windlass. Racing skippers claimed to be able to squeeze an extra half knot out of their steamboats by adjusting the shape of the hull by screwing or unscrewing the hogging-truss. The fact that the hulls of these steamers leaked, in consequence, even worse than the hulls of the triremes did not matter very much because they were provided with steam bilgepumps.

Trusses also occur, of course, in many different forms in connection with the rigs of almost every kind of sailing ship. Very probably, the sail is another Egyptian invention, for on the Nile the wind blows upstream for most of the year, so that cargo vessels can sail up the river with a fair wind and drift back downstream with the current – as they still do today.

The first problem in constructing a sailing ship is to erect some kind of mast upon which sail can be hoisted. The second, and much more difficult, problem is to keep that mast in place. Broadly speaking, the masts of conventional sailing ships are, structurally, simple poles or struts which are supported from a sufficient number of directions by the system of fixed ropes which seamen call 'standing rigging': that is to say, by 'shrouds' and 'stays'. If one has a hull which is rigid enough to withstand the pull of the shrouds and stays, this is nearly always the best arrangement, and (as we shall see in Chapter 14) it can be shown mathematically to minimize the weight and cost. However, the Egyptians had not done this sort of mathematics, and, furthermore, they had no preconceived ideas about the subject. All they knew was that they were rather tired of rowing and they wanted to find some way of supporting a new-fangled thing called a sail above a hull which was made from reeds.

Having spent a good deal of time in developing sailing rigs for the pneumatic rescue dinghies which were carried by bomber aircraft,* I can sympathize with the ancient Egyptians about this

* For the benefit of any unfortunate airmen who may have had involuntary experience of these devices, I would explain that I would go about the job quite differently nowadays.

business of masts. The blown-up hulls of the pneumatic dinghies were probably just about as flexible as the Egyptian reed boats. One cannot really expect to be able to attach highly-loaded ropes to a thing like a soggy balloon or to a floppy bundle of reeds, and in these circumstances the whole idea of 'standing rigging' becomes rather laughable. Very sensibly, therefore, the Egyptians merely planted a sort of tripod, or sometimes an 'A'-shaped truss, on top of the rather squidgy hull (Figure 10). This affair worked perfectly well on the Nile; I used to envy the ancient Egyptians their solution to the problem, which, unfortunately, was never practicable with the rescue dinghies. The Egyptians did not have to arrange for the whole of their sailing rig to be folded up and packed inside a small bag, which, in turn, had to be stowed in a crowded aircraft.

The hulls of Greek and Roman merchant ships were generally sufficiently strong and stiff to resist the loads imposed on them by conventional standing rigging, and so these vessels had their masts stepped in the middle of the ship and supported by shrouds and stays in the usual way. For some reason, however, even large Roman ships seldom got much beyond the stage of a single mast, carrying one large square sail, set from one long yard. It was not until the great expansion of sea voyaging at the time of the Renaissance that the rig of large sailing ships was elaborated by multiplying the number of masts and sails. About this time the single mast was replaced by three, called the fore, main and mizzen masts. Eventually, each of these masts was extended upwards so as to be able to carry, above the lower square sails or 'courses', first, square topsails, then topgallants, and finally royals. (The even loftier skysails and moonsails came much later, an affectation of the clipper era.)

Traditionally each sail – course, topsail, topgallant and royal – is set from its own separate section of mast. That is to say, each lower mast is surmounted by a topmast, each topmast in turn by a topgallant mast and so on. Each of these upper masts constitutes a separate piece of timber, and each is supported in its proper position by means of elaborate and sophisticated sliding fittings. These were arranged so that all the upper masts and yards could, on occasion, be lowered and sent down on deck. Since the larger spars each weighed several tons, it needed both skill and nerve to raise and lower such unwieldy objects in a rolling ship. However, a big warship would have a crew of 800 men, most of whom could have put both steeplejacks and trained athletes to shame. The sail-drill in the Mediterranean fleet in the 1840s has become legendary. It is alleged that, when the admiral had finished his breakfast, he was apt to signal 'All ships will strike topmasts. Report time taken and number of casualties'. However this may be, it is certain that crack battleships like H.M.S. *Marlborough* could be stripped to their lower masts by their own crews in a matter of minutes and re-rigged as quickly. These competitive exercises were by no means a waste of effort. Ships carried ample supplies of spare spars, and the safety of a ship in an emergency, or the outcome of an action in time of war, had repeatedly depended upon how quickly crippled masts could be replaced. A limited number of casualties during peace-time drills had to be accepted, as we accept accidents in riding or rock-climbing.

The structural technology behind all this was superb of its kind, and it is worthy of the attention of modern engineers, who are apt to be rather snooty about it. The complexity of the rigging which was needed to support all the tophamper in the later sailing ships can best be appreciated by going to look at the *Victory* (Plate 14) or the *Cutty Sark*. The total height of *Victory*'s mainmast, for instance, is about 223 feet (67 metres). The length of her main-yard is 102 feet (30 metres), but this can be extended at will to a total width of 197 feet (59 metres) by means of sliding stunsail booms. All this immense mechanism worked, and worked reliably, for years on end and in spite of the most appalling conditions of wind and sea, being much more reliable than most modern machinery.

The masts of big sailing ships represent perhaps the most elaborate and certainly one of the most beautiful systems of trussing which has ever been developed. At the cost of considerable complexity, the total weight of structure up aloft was kept down to a safe figure. However, when big guns, mounted in revolving turrets, had to be introduced into sailing battleships around 1870, the network of shrouds and other ropes was found to restrict unduly the arcs of fire of the guns. For this reason certain ironclads, notably H.M.S. *Captain*, were fitted with tripod masts which could be arranged so as to permit a better field of fire. This was a reversion to the Egyptian method of masting, if you like. However, the extra top-weight of these tripod structures had a bad effect upon the already precarious stability of these ships. This top-weight undoubtedly contributed to the capsizing of the *Captain*, under sail, one dirty night in the Bay of Biscay. Nearly five hundred men were drowned.

Cantilevers and 'simply supported' beams

It is evident that, functionally, it does not make much difference whether a 'beam' is in the form of a long continuous piece of material – a solid tree-trunk or a steel rod or tube or joist – or whether it takes the shape of some kind of open-work truss. This latter might be a wooden roof-truss, a sea-going arrangement of ropes and spars, or some modern Meccano-like lattice, such as a bridge or an electricity pylon. As we shall see, there are plenty of both kinds of beams in animals as well. The fact that bridges and roof-trusses and horses' backs and dachshunds are usually more or less horizontal, while ships' masts and telegraph poles and pylons and ostriches' necks are quite often vertical, does not make much difference. The essential purpose of all these structures is the same: that is to say, *a load which acts at right angles to the length of the beam is supported without putting any longitudinal force upon whatever is supporting the beam.* This is essentially what all beams are for.

It might be thought that a thing like a ship's mast was an exception to this, because a mast thrusts downwards, forcibly, upon the hull of a ship. But then the shrouds and stays pull upwards on the hull just as much, and so there is no *net* vertical force upon the hull, which does not rise or sink in the water in consequence. Similar arguments apply with many animal structures. A horse's neck, for instance, is very much like a mast. The vertebrae, like the mast, are in compression and push backwards on the horse's body, but they are stayed, like the mast, by the neck tendons, which pull forwards on the body with an equal and opposite force.

In the sense which we have just been discussing, all beams, living or dead, do the same job; yet beams as a whole tend to fall into two main categories: 'cantilevers' and 'simply supported' beams. There are in fact further variants and sub-divisions, which are frequently useful for examination and other purposes, but we shall ignore them for the moment.

Figure 11. A cantilever beam with distributed load.

A 'cantilever' is a beam one end of which can be considered as being 'built in' to some rigid support, such as a wall or the ground. This end-condition is called by engineers *encastré* – which is merely French for 'built in'. The other end of the cantilever, of course, sticks out and supports the load. Electricity pylons, tele-

Figure 12. Simply supported beam.

graph poles, ships' masts, turbine blades, horns, teeth, animals' necks and trees and cornstalks and dandelions are cantilevers, and so are the wings of birds and aeroplanes and butterflies and also the tails of mice and peacocks.

A simply supported beam (Figure 12) is one which rests freely on supports at both ends.

Structurally, the two cases are closely connected. From Figure 13 we can see that a simply supported beam is simply equivalent to two cantilevers, back to back and turned upside down.

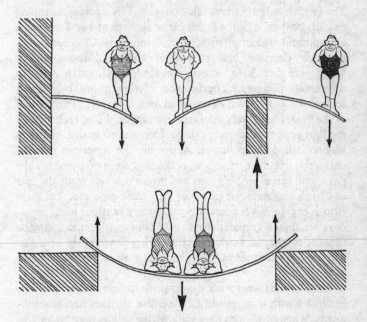

Figure 13. A simply supported beam may be considered as two cantilevers back to back and upside down.

Bridge trusses

The road is carried across valleys hundreds of feet in depth on rude trestle bridges, which creak and groan beneath the weight of the train. Anything apparently more insecure than these structures can hardly be found else-

where, and I always drew a long breath of relief as I found myself safely on the other side. It is a fearful thing to look out of the carriage windows into the dizzy depth below, and feel that if the frail fabric were to collapse, as it seemed on the point of doing, we should all be dashed to pieces with no possibility of escape. Even in the Eastern States many of these primitive bridges yet remain, and it is said that few accidents have happened from their use. They are, however, very liable to destruction from fire, caused by burning coals falling from the engine.

Rev. Samuel Manning, LL.D., *American Pictures* (1875)

The English railways were built straight and level across the rolling English landscape by the lavish use of cuttings and embankments and splendid viaducts of masonry and ironwork. All this engineering luxury depended upon supplies of capital and labour, both of which were plentiful in Victorian England. Conditions in America were totally different.* The distances were enormous; capital was scarce; the wages, even of unskilled men, were high. In the Land of the Free, where every man was an amateur, skilled craftsmen of the European type scarcely existed. Iron was expensive, but there was unlimited cheap timber. Above all, the American railroad engineers, like their steamboat colleagues, were prepared to take risks with other people's lives and property which made the hair of British engineers rise up under their stove-pipe hats. Yet these British engineers were certainly not unduly cautious men; nowadays we should consider them rash. Nineteenth-century Americans, of course, were in the habit of living dangerously – but this was more on account of their engineers than of the Red Indians or the bandits.

The railroads were pushed westwards as fast as they could be built and with a minimum of expensive cuttings and embankments. When conditions were suitable, the valleys were bridged by means of those enormous timber trestle viaducts which alarmed the Rev. Dr Manning. They will always be associated, in tradition, with the American railways; a fair number of them survive today (Plate 15). Once they had been constructed, the American railways were vastly profitable – the Central Pacific Railroad is said

---

*The cost per mile of American railways was one fifth of that of English lines, although American wages were much higher.

to have paid dividends of 60 per cent – and so they were soon able to convert many of their precarious trestle bridges to solid earth embankments by tipping soil from the top from specially constructed trains until the whole wooden structure was encased in earth and could be left to rot away.

Wide and rolling rivers could not be crossed by the trestle viaducts and so there was a need for large, long-span bridges. Permanent bridges of the European type were often impracticable for lack of money and skilled labour, and so there was a very active requirement for long – and cheap – wooden trusses, which could be made by ordinary joiners. Since the construction of these trusses was potentially profitable and since the Americans are an incurably inventive people, a very considerable number of nineteenth-century Americans seem to have spent their time in inventing trusses. There are therefore to be found in the text-books a very considerable number of designs for bridge trusses, each slightly different, and each called after the name of its inventor. We need not go through them all in detail, for they all work upon somewhat similar principles, but two or three types are worth mentioning.

One of the earliest of these was the Bollman truss (Figure 14),

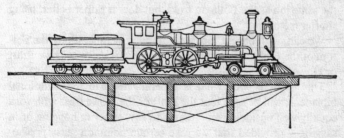

Figure 14. Bollman truss.

which was very extensively used in America – perhaps more on account of Bollman's political talents than his technical ones. He somehow managed to persuade the American government that his was the only 'safe' design of truss, and at one time its use was made compulsory. This may not have been quite so difficult a legislative feat as one might suppose, since it came to be accepted for many years as a practical working principle, by professional engineers, that the technical ignorance of the American Congressman could safely be regarded as bottomless.*

Figure 14 shows a simplified Bollman truss with only three panels. In practice there were usually a great many more, and the whole thing tended to get complicated. Besides this the tension members were unnecessarily long. The Fink truss (Figure 15) does

Figure 15.   Fink truss.

the same job as the Bollman truss, but does it rather better, using shorter members.

We can, with benefit, put a continuous member along the bottom of the Fink truss and turn it into what is more or less a Pratt or Howe truss (Figure 16).

This is pretty well what is generally used in the traditional biplane. It will be seen that the Pratt or the Howe truss will work equally well upside-down – that is to say, either in hogging or in sagging – provided that we take certain common-sense precau-

---

*As late as 1912, during the American governmental inquiry into the loss of the liner *Titanic*, the following exchange was recorded:
Senator X.: You have told us that the ship was fitted with watertight compartments?
Expert witness: Yes.
Senator X.: Then will you explain how it was that the passengers were not able to get inside the watertight compartments when the ship sank?

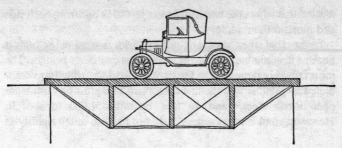

Figure 16.   Pratt or Howe truss.

tions. Furthermore, if we arrange that all the members can take both tension and compression, we can simplify the structure by turning it into a Warren girder (Figure 17). It is this form, or something like it, which is most commonly used for trusses made from ordinary steelwork.

So far, we have considered all these bridges as being simply supported beams, and so, of course, a great many of them were and are. However, a number of beam bridges are cantilever bridges. For some reason cantilever bridges were never very

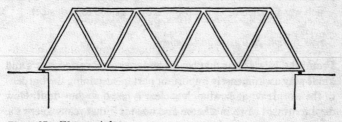

Figure 17.   Warren girder.

popular in wooden construction, but they are widely used nowadays when built from steel and concrete. A good proportion of the bridges over the motorways are reinforced concrete cantilever bridges. Such bridges generally have a centre-section which is a simply supported beam, resting on the extremities of two cantilevers (Figure 18). This is partly because it is easier to accommodate the deflections with this arrangement. However, there are a few bridges where the two cantilevers just stick out from each side and meet in the middle.

In the days when very long railway bridges were being built it became fashionable to construct large steel cantilever bridges. The most famous example is the Forth railway bridge, which was completed in 1890. It was the first important bridge to be built from open-hearth steel,* and, in fact, contains 51,000 tons of it. However, road bridges generally do not need so much rigidity as

Figure 18.   Cantilever bridge with simply supported beam for centre section.

railway bridges (the Forth bridge is said to be the only large bridge in the world over which trains are allowed to pass at full speed), and so most long modern bridges are suspension bridges, which are usually cheaper to build. The Forth road bridge, which has a similar total span to the railway bridge next door to it, and which was finished in 1965, contains only 22,000 tons of steel.

Stress systems in trusses and beams

From all this it is clear that beams and trusses of various sorts and kinds play an immensely important part in sustaining the burdens of the world. What is rather less clear is just how they do it. How do the stresses work in a beam and what is it that really keeps the thing up? As we have said, lattice trusses and solid beams can nearly always be used interchangeably, and so, as one might suppose, the stress system within a truss is not very different in principle from that in a solid beam, although it has the advantage of being rather easier to visualize. Furthermore, cantilevers are perhaps easier to think about than simply supported beams, although as we have seen from Figure 13, the two conditions are quite simply related.

The New Science of Strong Materials, Chapter 10.

Let us consider therefore a truss in the form of a cantilever which is fixed to a wall (or *encastré*) at one end and which sticks out and supports a load *W*, for instance, from the other end. Let us begin, in fact, with the embryonic or nascent cantilever which is the simple triangular arrangement shown in Figure 19. In this

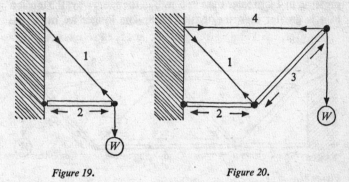

Figure 19.   Figure 20.

affair the weight, *W*, is directly kept from falling down by the action of the upward component of the tension in the slanting member 1. The compressive force in the horizontal member 2 can only act horizontally, and so it can play no *direct* part in sustaining the weight. However, they also serve who only push horizontally, and member No. 2 is performing an indirect but very necessary function in keeping the truss extended, that is to say, sticking out in the way it does.

Let us now add an extra panel to the truss, as in Figure 20. It is clear that the weight is now sustained *directly* by the *combined* upward action of the tension in No. 1 and the compression in No. 3. No. 4 is necessarily in tension but, like No. 2 (which is still in compression), it does not contribute directly to sustaining the weight, although the truss cannot hold up without it.

If we build the truss up into several panels, as in Figure 21, the general situation remains very much the same. The diagonal members 1 and 5 are in tension and 3 and 7 are in compression. It is still these members which directly sustain the load. Taken together, these members are resisting what is called 'shear'. We shall have a good deal more to say about shear in the next chapter.

In the meantime we may observe that the force which is acting in all of these diagonal members is numerically similar. This remains true however long the cantilever may be and however many panels it has.

This is not true, however, of the horizontal forces. The compression in 2 is greater than in 6 and, in the same way, the tension in 4 is greater than the tension in 8. The longer we make the

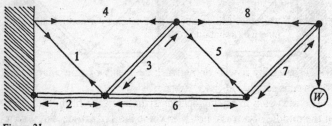

Figure 21.

cantilever, the higher the compression will be in member No. 2 and the greater the tension in No. 4. If we make the cantilever very long, then the horizontal or longitudinal tension and compression forces and stresses close to the fixed end may be very high indeed. In other words, such a cantilever will probably break near its root, which after all is only common sense. However, we do have the apparent paradox that the forces are highest in members which do not contribute directly to supporting the load.

In Figure 21 the downward load, or 'shearing force', is directly supported, as we said, by the zig-zag of the diagonal members 1, 3, 5 and 7. However, there is nothing to prevent us from complicating this diagonal trellis by introducing more slanting members, which will all perform the same function. In fact this is often done for various reasons (Figure 22). This is just what Nature quite frequently does. The trunk and rib-cage of most vertebrates can be considered as a sort of simply supported beam. This is obvious in the case of a horse. The bones of the vertebrae and the ribs form the compression members of a rather elaborate Fink truss (Figures 15 and 23). The space between the ribs is

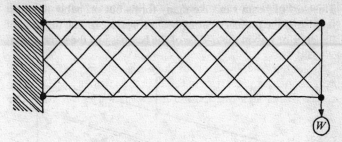

Figure 22. The shear can equally well be taken by a multiple lattice or indeed by a continuous plate.

criss-crossed by a web or network or trellis of muscular tissue which runs roughly at $\pm 45°$ to the ribs.

The next step in an engineering structure is to fill in the space in the middle of a truss, not with some kind of lattice, but with a continuous plate or 'web' of some material like steel or plywood.

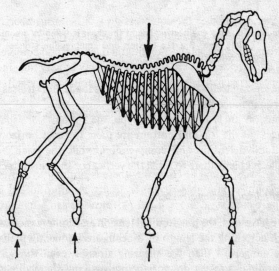

Figure 23. Many vertebrate animals form a sort of Fink truss with muscles and tendons making a rather complicated diagonal shear bracing between the ribs.

This sort of beam can take many forms but probably the most familiar is the ordinary H or I beam (Figure 24). The function of the plate or web in the middle of the beam is just the same as that

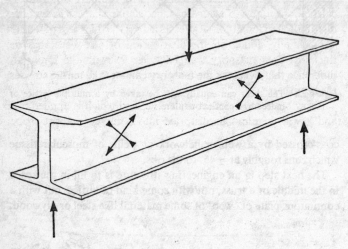

Figure 24. In many engineering beams the shear is taken by a continuous plate web. But the tension and compression stresses due to shearing are still at ±45°.

of the zig-zag trellis in a truss, and so the loads and stresses in the web run in much the same way.

Thus, in an H beam of this type, the 'booms' or 'spars' or 'flanges' at the top and bottom are there to resist horizontal or longitudinal tension and compression, while the 'web', in the middle, is chiefly there to resist the vertical or shearing forces.

Longitudinal bending stresses

As we have said, the longitudinal tension and compression stresses which act along the length of a beam are frequently higher and more dangerous than the shearing stresses, even though these longitudinal stresses do not themselves contribute directly towards supporting the load. In the ordinary beams which we are likely to meet in practice, it is very commonly the longitudinal stresses which are liable to cause failure, and so they are frequently the first stresses to be calculated by an engineer.

Although beams of H section (Figure 24) are common, a beam may be of any cross-sectional shape, and ordinary beam-theory calculations apply to beams of most simple shapes. In fact, the distribution of longitudinal stresses across the thickness of a beam is essentially similar to the distribution of stresses across the thickness of a masonry wall (Chapter 9), with the important difference that, whereas the masonry cannot take tensile stresses, the beam can.

Every beam must deflect under the load which is applied to it and it will therefore be distorted into a curved or bent shape.

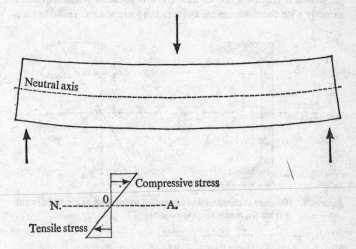

Figure 25.   Distribution of stress through the thickness of a beam.

Material on the concave or compression face of a bent beam will be shortened or strained in compression. Material on the convex or tension face will be lengthened or strained in tension (Figure 25). If the material of the beam obeys Hooke's law the distribution of stress and strain across any section of the beam will be a straight line, and there will be some point '0' at which the longitudinal stress and strain is neither tensile nor compressive, but is zero.

This point lies on what is called the 'neutral axis' (N.A.) of the beam.

Since it is important to know the position of the neutral axis in a beam it is fortunate that this is easy to determine. It is quite simple to show, algebraically, that the neutral axis must always pass through the centroid or 'centre of gravity' of the cross-section of the beam. For simple symmetrical sections, such as rectangles and circles and tubes and H beams, the neutral axis lies in the middle, half-way between the top and bottom of the beam. For non-symmetrical sections, such as railway lines and ships and aircraft wings, its position will have to be calculated – but this is not very difficult.

It is clear from Figure 25 that the longitudinal stress increases directly with the distance away from the neutral axis. This distance

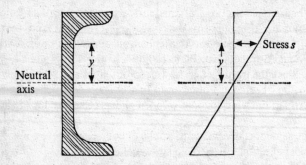

Figure 26. Tension or compression stress due to bending at a point distant *y* from the neutral axis is *s* where

$$s = \frac{My}{I}$$

and M = bending moment
I = second moment of area of cross-section.
For how to arrive at M and I, see Appendix 2.

is generally called *y* when discussing beam theory.* Now if we are seeking structural 'efficiency', whether in terms of weight of material, or cost, or metabolic energy, then we do not want to keep any cats that don't catch mice. In other words we do not

*See Appendix 2.

want to have to provide material which carries little or no stress. This means that we want, as far as possible, to discard material which lies close to the neutral axis in favour of material as far away from it as possible. Of course, we shall need to leave some material near the neutral axis so as to carry the shearing stresses, but in practice we may not need much material for this purpose and quite a thin web may suffice (Figure 26).

This is why, in engineering, steel beams usually have a cross-section of H or 'channel' or Z form (Figure 24). These sections have the advantage of being relatively easy to make from mild steel in a rolling-mill. They are often known as 'rolled steel joists' (R.S.J.s), and nowadays they can be bought in very large sizes. Z sections have the advantage over channels and Hs that it is easier to rivet the flanges to a plate. This is why Zs are often used for ships' ribs.

When simple sections of this sort are unsuitable it is quite common to use built-up 'box' sections. The first and most important use of these was in Stephenson's Britannia bridge over the Menai Straits (1850; Plate 16 and Chapter 13, Figure 11, p. 291). Since the introduction of waterproof glues and reliable plywood, box beams are often used in wooden construction, particularly in the wing-spars of wooden gliders (Chapter 13, Figure 5, p. 279).

The same sort of arguments apply, of course, when we come to consider sheet materials. Thin sheet metal is weak and flexible in bending, and, to save weight, we want, if possible, to achieve a deeper section. This is often done by rolling corrugations into the metal sheet – with corrugated iron as the unfortunate result.* Corrugated metal sheet has been used in the past for the outside skins of both ships and aircraft, notably with the old Junkers monoplanes. However, the objections are obvious, and it is much more usual nowadays to stiffen and strengthen metal skins in shipbuilding and in aerospace by riveting or welding metal angles, called stringers, to the inside surfaces of the skin.

In all these situations the load commonly comes upon the beam from one direction only, and the shape of the cross-section is

* Notice also the corrugations in clam-shells and in many kinds of leaves, such as hornbeam.

optimized with regard to this condition. In some engineering structures and in very many biological ones, however, the load may come from any direction. This is roughly true for lamp-posts, chair legs, bamboos and leg-bones. For such purposes it is better to use a round, hollow tube, and of course this is what is very often done. An intermediate case occurs with bermuda masts. These are generally made from tubes of oval or pear-shaped section. This is not primarily so as to reduce wind-drag by 'streamlining', as is often supposed, but rather to cater for the fact that it is much easier to stay a modern mast laterally than it is in the fore and aft plane, and so the mast section has to take account of this by providing more strength and stiffness fore and aft.

Chapter 12 The mysteries of shear and torsion

– or Polaris and the bias-cut nightie

> *Twist ye, twine ye! even so*
> *Mingle shades of joy and woe,*
> *Hope and fear, and peace and strife,*
> *In the thread of human life.*

Sir Walter Scott, *Guy Mannering*

There is supposed to have been a book review by Dorothy Parker which started off 'This book tells me more than I care to know about the Principles of Accountancy'. And indeed I dare say that many of us are apt to come to the conclusion that the way in which things behave in shear might, after all, be left to the experts. Tension and compression we feel we can cope with, but when it comes to shear we think we can detect a tendency for the mind to boggle.

It is unfortunate, therefore, that the shear stresses to which we are introduced in the elasticity text-books are assumed to spend their time inhabiting things like crankshafts or the more boring sorts of beams. Though undeniably worthy, this approach somehow lacks human appeal, and it also diverts attention from the fact that shearing stresses and shearing strains are by no means confined to beams and crankshafts but keep intruding into practically everything we do – sometimes with unexpected results. This is why boats leak, tables wobble and clothes bulge in the wrong places. Not only engineers, but also biologists and surgeons and dressmakers and amateur carpenters and the people who make loose covers for chairs would live better and more fruitful lives if they could only look a shear stress between the eyes without flinching.

If tension is about pulling and compression is about pushing, then shear is about sliding. In other words, a shear stress measures the tendency for one part of a solid to slide past the next bit: the sort of thing which happens when you throw a pack of cards on the table or jerk the rug from under someone's feet. It also nearly always occurs when anything is twisted, such as one's ankle or the driving shaft of a car or any other piece of machinery. Materials which are being sheared or twisted usually behave in quite straightforward and rational ways, but, rather naturally, when we come to discuss this behaviour it helps a good deal to make use of the appropriate vocabulary. So we might begin with a few definitions.

The vocabulary of shear

The elasticity of shear is very much like the elasticity of tension and compression, and concepts like shear stress, shear strain and shear modulus are pretty closely analogous to their tensile equivalents and certainly no harder to understand.

SHEAR STRESS – N

As we have said, a shear stress is a measure of the tendency for one part of a solid to slide past the neighbouring part, very much

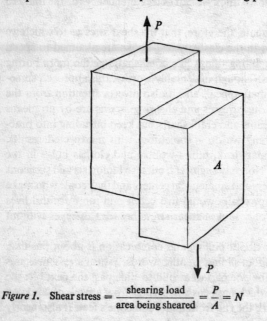

Figure 1. Shear stress $= \dfrac{\text{shearing load}}{\text{area being sheared}} = \dfrac{P}{A} = N$

as in Figure 1. Hence, if a cross-section of material, having an area A, is acted upon by a shearing force P, then the shear stress in the material at that point will be

$$\text{shear stress} = \frac{\text{shearing load}}{\text{area being sheared}} = \frac{P}{A} = N, \text{ let us say}$$

– just like a tensile stress. The units are also the same as those of a tensile stress, that is to say, p.s.i., MN/m² or what you fancy.

SHEAR STRAIN – g

All solids yield or strain under the action of a shear stress, in the same sort of way as they do under a tensile stress. In the case of shear, however, the strain is an angular one, and it is therefore

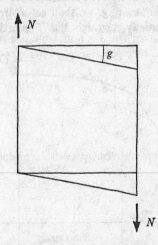

Figure 2. Shear strain = angle through which material is distorted as a result of shear stress N
= g, which is an *angle* – usually in radians.

measured, like any other angle, in degrees or in radians – usually in radians (Figure 2). Radians, of course, have no dimensions, being really a number or a fraction or a ratio. We shall call the shear strain g in this book: like the tensile strain, e, therefore, g is a dimensionless number or fraction and has no units.

In hard solids like metal or concrete or bone, the elastic shearing strain is likely to be less than 1° (1/57 radian). Beyond this shearing strain, materials of this kind will generally either break or else flow in a plastic and irrecoverable way, like butter. However, with materials like rubber or textiles or biological soft tissues, recoverable or elastic shear strains may be much higher than this – perhaps 30° to 40°. With liquids and squidgy things like treacle or custard or plasticine, the shear strain is unlimited; but then it is not recoverable.

THE SHEAR MODULUS OR MODULUS OF RIGIDITY – G

At small and moderate stresses most solids obey Hooke's law in shear, much as they do in tension. Thus, if we plot the shear stress, N, against the shear strain, g, we shall get a stress-strain curve which is, at least initially, a straight line (Figure 3). The slope or

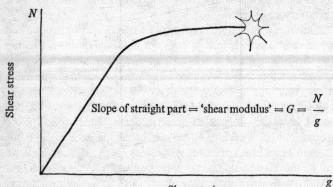

Slope of straight part = 'shear modulus' = $G = \dfrac{N}{g}$

Figure 3. The stress-strain diagram in shear is very like that in tension. The slope of the straight part is equivalent to the shear modulus

$$G = \frac{N}{g}$$

gradient of the straight part represents the stiffness of the material in shear and is called the 'shear modulus', or sometimes the 'modulus of rigidity', or 'G'. Thus

$$\text{shear modulus} = \frac{\text{shear stress}}{\text{shear strain}} = \frac{N}{g} = G*$$

So G is the exact analogue of the Young's modulus, E, and, like E, it has the dimensions and units of a stress: that is to say, p.s.i., MN/m^2 or whatever.

Shear webs – isotropic and anisotropic materials

As we said in the last chapter, although there may be large horizontal tension and compression forces in the top and bottom flanges of a beam or a truss, the actual upward thrust which really enables the structure to do its job of sustaining a downward load has to be produced by the web – that is to say, by the part in the middle which joins the top and bottom booms together. In a continuous beam the web will be of solid material, perhaps a metal plate; in a truss the same function will be served by some sort of lattice or trellis.

Since the distinction between a material and a structure is never very clearly defined, it does not matter very much whether the shearing loads in a beam are carried by a continuous plate web or whether they are carried by a lattice which might be made up of rods and wires, strips of wood or whatever. There is, however, an important difference. If the web is made from, say, a metal plate, then it is of no consequence in which direction the plate is put on. That is to say, if we cut the plate for the web out of some larger sheet of metal, it does not matter at what angle we cut it, since the metal has the same properties in every direction within itself. Such materials, which include the metals, brick, concrete, glass and most kinds of stone, are called 'isotropic', which is Greek for 'the same in all directions'. The fact that metals are isotropic (or nearly so) and have the same properties in

* Note that there is a relationship between G and E. For isotropic materials like metals

$$G = \frac{E}{2(1+q)}$$

where q = Poisson's ratio.

all directions makes life somewhat easier for engineers and is one of the reasons why they like metal.

However, if we now consider the lattice web, it is clear that it must be constructed so that the rods and tie-bars lie nearly at ±45° to the length of the beam. If this is not done, then the web will have little or no stiffness in shear (Figures 4–5). Under load the lattice will fold up and the beam will probably collapse. Materials of this kind are called 'anisotropic', or sometimes

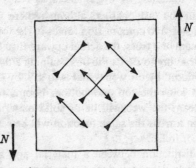

Figure 4. Shear will produce tension and compression stresses in directions at 45° to the plane of shearing.

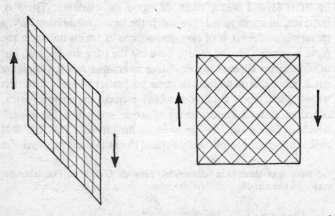

Figure 5. Thus a system like the one on the right is 'rigid' in shear, and systems like the one on the left are floppy.

'aelotropic' – both of which are Greek for 'different in different directions'. In their different ways wood and cloth and nearly all biological materials are anisotropic and they tend to make life complicated, not only for engineers, but for a great many other people as well.

Cloth is one of the commonest of all artificial materials and it is highly anisotropic. As we have said repeatedly, the distinction between a material and a structure is a vague one, and cloth, though called 'material' by dressmakers, is really a structure, made up of separate yarns or threads crossing each other at right angles; and its behaviour under load is much the same as that of the trellis web of a beam or a truss.

If you take a square of ordinary cloth in your hands – a handkerchief might do – it is easy to see that the way in which it deforms under a tensile load depends markedly upon the direction in which you pull it. If you pull, fairly precisely, along either the warp or the weft threads,* the cloth will extend very little; in

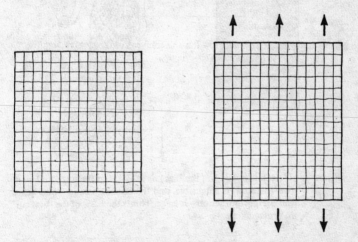

Figure 6. When cloth is pulled parallel to the warp or the weft threads, the 'material' is 'stiff' and the lateral contraction is quite small.

*Warp threads or yarns are those which run parallel to the length of a roll of cloth; weft threads are those which run across the cloth, at right angles to its length.

other words, it is stiff in tension. Furthermore, in this case, if one looks carefully, one can see that there is not much lateral contraction as a result of the pull (Figure 6). Thus the Poisson's ratio (which we discussed in Chapter 8 in connection with arteries) is low.

However, if you now pull the cloth at 45° to the direction of the threads – as a dressmaker would say, 'in the bias direction' – it is much more extensible; that is to say, Young's modulus in tension is low. This time, though, there is a large lateral contraction, so

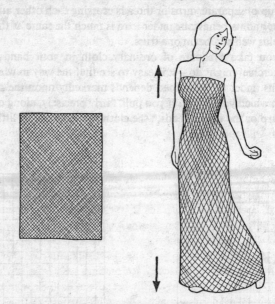

Figure 7. If cloth is pulled 'on the bias' or at ±45° to the warp and weft, the 'material' is extensible, and the Poisson's ratio – and hence the lateral contraction – is large. This is the basis of the 'bias cut' in dressmaking.

that, in this direction, the Poisson's ratio is high; in fact it may have a value of about 1·0 (Figure 7). On the whole, the more loosely the cloth is woven, the greater is likely to be the difference between its behaviour in the bias and in the warp and weft or 'square' direction.

Although I suppose that not very many people have ever heard of the word 'anisotropy', the fact that cloth behaves in this sort of way must have been familiar to nearly everybody for centuries. Rather surprisingly, however, the technical and social consequences of the anisotropy of woven cloth do not seem to have been properly realized or exploited until quite recent times.

When we stop to think about the matter, it is clear that when we make anything from cloth or canvas, we can minimize the distortions by arranging for the important stresses to run, as far as possible, along the directions of the warp and weft threads. This usually involves cutting the material 'on the square'. If the circumstances are such that the cloth is pulled at 45°, that is to say 'on the bias', then we shall get much larger distortions, which will, however, be symmetrical. But, should we be so inept that the cloth ends up by being pulled in some intermediate direction, which is neither one thing nor the other, then we shall not only get large distortions, but these will be highly asymmetrical. Thus the cloth will pull into some weird and almost certainly unwelcome shape.*

Although sailmaking has been an important industry ever since the beginning of history, these elementary facts about canvas never fully dawned upon European sailmakers. They continued from age to age to construct sails in such a way that the pull came obliquely upon both the warp and weft threads. As a consequence, their sails quickly became baggy and could seldom be made to set properly when the wind was ahead. The situation was worsened by the European predilection for making sails from flax canvas, which distorts particularly easily because of its loose weave.

Rational modern sailmaking began in the United States early in the nineteenth century. American sailmakers used tightly woven cotton canvas, and they arranged their seams in such a way that the direction of the threads corresponded more nearly to the direction of the applied stresses. Although the consequence was that American ships could frequently sail faster and also closer to

* An understanding of this principle is very important when making things like balloons and pneumatic dinghies from rubberized fabric. If shear distortions are incurred the rubber coating is strained in such a way that the fabric will leak.

the wind than British ones, it required something like an earth-quake to bring the facts home to English sailmakers. This was provided by the publicity associated with the schooner yacht *America*, which came over from New York to Cowes in 1851 to compete with the fastest English yachts. She was entered for a race round the Isle of Wight which was to be sailed for a rather ugly piece of silverware presented by Queen Victoria. This jug-like object has since acquired a certain fame as the 'America's

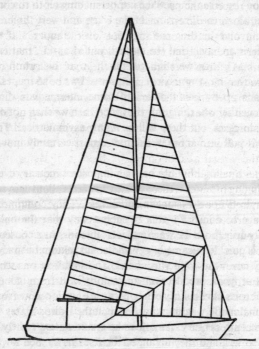

Figure 8. In modern sailmaking it is usual to arrange the weft threads of the canvas so that they are parallel to the free edges of the sail.

Cup'. When the Queen was told that the *America* was the first yacht to have crossed the finishing line, she asked 'And who is second?'

'There is no second in sight yet, your majesty.'

After this, the English sailmakers mended their ways – so much so that, within a few years, American yachtsmen would be buying their sails from Mr Ratsey of Cowes. The lessons taught by the American sailmakers have stuck, and, although the majority of modern sails are made from Terylene, not cotton, if you look at any modern sail (Figure 8) you can see that it is cut in such a way that the weft threads are, as far as possible, parallel to the free edges of the sail, which is usually the direction of greatest stress.

In many respects the problems of persuading cloth to conform to a desired three-dimensional shape are not very different in sailmaking and in dressmaking. However, tailors and dressmakers seem to have been more intelligent about the matter than sailmakers. As far as was practicable they cut their cloth on the square, so that most of the circumferential or hoop stresses came directly along the line of the yarns. When a close fit was wanted it was achieved by what might be described as a system of Applied Tension: in other words, by lacing. At times the Victorian young lady seems to have had nearly as much rigging as a sailing ship.

With the virtual abandonment of systems of lacing in post-Edwardian times – possibly on account of a shortage of ladies' maids – women might well have had to face a shapeless future. However, in 1922 a dressmaker called Mlle Vionnet set up shop in Paris and proceeded to invent the 'bias cut'. Mlle Vionnet had probably never heard of her distinguished compatriot S. D. Poisson – still less of his ratio – but she realized intuitively that there are more ways of getting a fit than by pulling on strings or straining at hooks and eyes. The cloth of a dress is subject to vertical tensile stresses both from its own weight and from the movements of the wearer; and if the cloth is disposed at 45° to this vertical stress one can exploit the resulting large lateral contraction so as to get a clinging effect. The result was no doubt cheaper and more comfortable than the Edwardian solutions to the problem and, in selected instances, probably more devastating (Plates 17 and 18).

An analogous problem arises with the design of large rockets. Some rockets are driven by combinations of liquid fuels such as kerosene and liquid oxygen, but these systems involve elaborate plumbing which is liable to go wrong. Thus it may be better to use a 'solid' fuel such as that known as 'plastic propellant'. This stuff burns vigorously but relatively slowly, producing a great volume of hot gas which escapes through the rocket nozzle with a most impressive noise, driving the thing along as it does so. Both the propellant and the gas which it produces are contained within a strong cylindrical case or pressure vessel, whose walls must not be unduly exposed to flames or to high temperatures. For this reason the rather massive propellant charge is shaped in the form of a thick tube which fits tightly into the rocket casing. When the rocket is fired, combustion takes place at the inner surface of the plastic propellant, so that the tubular charge burns from the inside outwards. In this way the material of the case is protected from the flames up to the last possible moment by the presence of the remaining unburnt fuel.

Plastic propellant looks and feels rather like plasticine, and, like plasticine, it is apt to break in a brittle way, especially when it is cold. When a rocket is firing, the case naturally tends to expand under the gas pressure, rather as an artery expands under blood-pressure; if it does so, then the propellant has to expand with it. If the interior of the charge is still cold, it is likely to crack when the circumferential strain in the case reaches about 1·0 per cent. If this happens, then the flames will penetrate down the crack and destroy the case. This naturally results in a sensational explosion as another Polaris bites the dust.

Round about 1950, it occurred to some of us that it would be advantageous to make the rocket case, not from a metal tube, but in the form of a cylindrical vessel, wound from a double helix of strong glass fibres, bonded together with a resin adhesive. If the fibre angles are calculated correctly, it is possible so to arrange things that the change of diameter of the tube under pressure is small. It is true that, in such a situation, the tube will elongate more than it otherwise would, like Mlle Vionnet's waists, but, for various reasons, a longitudinal extension is less damaging to the propellant. As I seem to remember, this idea about rockets stemmed from the bias-cut nighties which were around at the time.

The strain requirements for rockets are generally just the opposite of what is needed in blood-vessels. As we saw in Chapter 8, one wants an artery to maintain a constant length while exposed to fluctuations in blood-pressure (but changes in artery diameter are not important). Either condition can be met by making suitably designed tubes from helically disposed fibres. Problems of this kind keep cropping up in biology, and it was most interesting to find that Professor Steve Wainwright of Duke University, who is concerned with worms, has derived, quite independently, just the same mathematics as we had worked out twenty years or so before for use in rocketry.* On inquiry, I find that in this case too the inspiration arose, via Professor Biggs, from the bias cut.

The invention of the bias cut brought fame to Mlle Vionnet in the world of haute couture. She lived to a great age and died, not long ago, at ninety-eight, quite unaware of her very significant contributions to space travel, to military technology and to the biomechanics of worms.

Shear stress is only tension and compression acting at ±45° – and vice versa

A very little further thought about plate webs in beams and lattice webs in trusses and about bias-cut nighties makes it obvious that a shear stress is merely tension or compression (or both) acting at 45°, and that, furthermore, there is a shear stress acting at 45° to every tension and compression stress.

In fact solids, especially metals, very frequently break in tension by reason of the shear stress at 45°. It is this which leads to the 'necking' of metal rods and plates in tension and to the mechanics of ductility in metals (Figure 9 and Chapter 5).

As we shall see in the next chapter, very much the same thing can also occur in compression. That is to say, many solids break in compression by sliding away from the load in shear.

Creasing – or the Wagner tension field

A thick plate or a solid piece of metal is able to resist compression, and so, when such things are subjected to shearing loads, there

*The cuticles of many worms and other soft animals are strengthened by systems of helically disposed collagen fibres (Chapter 8). The worm has much the same problems as the dressmaker, though it is often more successful in solving them. It is difficult to put a crease into a worm.

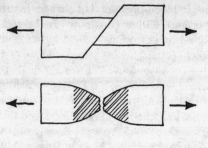

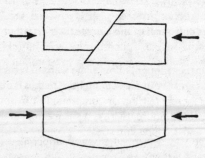

Figure 9. In ductile materials both tension and compression failure tend to occur by shear.

will exist, at $\pm 45°$, both tension and compressive stresses. Thin panels and membranes and films and fabrics are scarcely able to resist compression forces in their own plane, and so, when they are sheared, they are apt to crease. This creasing in shear is quite common in thin metal panels, such as occur in aircraft, and it is quite usual to see a creased or quilted effect on the surfaces of wings and fuselages due to this cause (Plate 19). This is called by engineers a 'Wagner tension field'.

The same effect is even more common in clothes and loose covers and tablecloths and badly cut sails. I suppose dressmakers do not very often talk about Wagner tension fields, but they do sometimes refer to that slightly mysterious quality which is known in the textile trade as 'drape'. The drape of a fabric depends mainly upon its shear modulus, and although, very probably, few couturiers could quote any figures – in SI or any other units – for the shear modulus, G, of their silks and cottons, on the whole, the lower the shear modulus of a 'material', the less its tendency to unwanted creasing. The reason why we cannot dress ourselves in paper or Cellophane without appearing ridiculous is mainly that these substances have too high a shear stiffness, so that they will not drape properly. Contrariwise, knitted and crêped fabrics have both a low Young's modulus and a low shear modulus, so that it is easy to get a close and flexible fit – as girls have discovered with knitted sweaters. In the same way the skin of young people has a low initial Young's modulus and a low shear modulus and there-fore conforms easily to the shape of the body.* In later life the skin becomes stiffer in shear, with obvious results. Recently Professor R. M. Kenedi of the University of Strathclyde has made an extensive study of elastic conformity in human skin. So, for the first time, the wrinkles of age are likely to be put on to a numerical or quantitative basis.

Torsion or twisting

The aeroplane was developed from an impossible object into a serious military weapon in something like ten years. This was achieved almost without benefit of science. The aircraft pioneers were often gifted amateurs and great sportsmen, but very few of them had much theoretical knowledge. Like modern car enthu-siasts, they were generally more interested in their noisy and unreliable engines than they were in the supporting structure, about which they knew little and often cared less. Naturally, if you hot up the engine sufficiently, you can get almost any aero-plane into the air. Whether it stays there depends upon problems of control and stability and structural strength which are con-ceptually difficult.

* Note that, for an initially flat membrane to conform easily to a surface with pronounced two-dimensional curvature, it is necessary to have *both* a low Young's modulus *and* a low shear modulus. This is essentially the prob-lem of map-projection which was encountered by Mercator about 1560.

In the early days too many brave men, like C. S. Rolls and S. F. Cody, paid with their lives for this attitude of mind. The theoretical basis of aerodynamics had been worked out by F. W. Lanchester in the 1890s, but not many practical men had the least idea what he was talking about.* A good many of the accidents to the pioneers were caused by stalls and spins, but structural failures were nearly as common. Since the early pilots seldom used parachutes, these accidents were generally fatal.

The requirement for a really reliable lightweight engineering structure was, of course, more or less a new one. In the first place, the wings of an aircraft are subject to bending forces, very much like a bridge. Since this is obvious, and since there was a good deal of precedent to go on in the matter of bridge construction, bending loads could generally be dealt with more or less safely. What was not so often realized was that the wings of an aeroplane are, in addition, subject to large torsional or twisting forces. If no proper provision is made to resist these torsions, the wings will be twisted off.

With the expansion of military flying after war broke out in 1914, the accident rate became a serious matter. In this country, luckily, such questions were dealt with by that small group of brilliant young men at Farnborough who afterwards became famous as Lord Cherwell, Sir Geoffrey Taylor, Sir Henry Tizard and 'Jehovah' Green. Thanks to their efforts the traditional biplane became, by 1918, one of the safest of all structures and came to be regarded as almost unbreakable. The Germans were less fortunate. Their aircraft technical authorities at that period had the reputation of being rather hidebound. At any rate they had a long run of structural accidents – many of them due to a failure to understand the problem of torsion in aircraft wings.

By the early part of 1917 the Allies had achieved a degree of

---

* Nor had many of the academic engineers. Even as late as 1936, the basic Lanchester-Prandtl (or vortex) theory of fluid dynamics was neither taught nor permitted to be used in the Department of Naval Architecture in the University of Glasgow. To those of a younger generation who may not be disposed to believe this story, I would point out that (a) I was myself a student in the department at the time, and (b) much the same sort of thing happens with 'modern' theories of fracture mechanics (Chapter 5) in present-day engineering departments.

air superiority on the western front, partly as a result of the technical quality of their fighters. However, in the meantime, the very able designer Antony Fokker was developing an advanced monoplane fighter – the Fokker D8 – with a performance better than anything available or in immediate prospect on the Allied side. Because of the critical tactical situation, production of the D8 was accelerated and it was issued to several of the crack German fighter squadrons without undergoing any adequate programme of test flying.

As soon as the D8 was flown under combat conditions it was found that, when the aircraft was pulled out of a dive in a dogfight, the wings came off. Since many lives were lost – including those of some of the best and most experienced German fighter pilots – this was a matter of very grave concern to the Germans at the time, and it is still instructive to study the cause of the trouble.

In those days most aircraft were biplanes, because this form of construction was lighter and also more reliable. However, for a given engine power, a monoplane will generally be faster than a biplane, because it does not have to experience the extra air resistance resulting from the aerodynamic interference which occurs between two adjacent sets of wings. There was thus a strong inducement to build monoplane fighters. However, although the reasons for the many failures were not understood, monoplanes had been known to be structurally unreliable ever since the wings of Samuel Langley's historic aeroplane had collapsed over the Potomac river in America in 1903.

The wings of the Fokker D8, like those of most monoplanes at the time, were fabric-covered. The fabric was there solely to provide the desired aerodynamic shape. It was merely stretched over an internal structural framework and itself carried none of the main loads. The main bending loads were taken by two parallel wooden spars or cantilever beams which projected sideways from the fuselage. The two spars were connected every few inches by a series of light shaped wooden ribs, to which the doped fabric was attached (Figure 10).

As soon as the accidents to the D8 became known the German Air Force authorities very naturally ordered structural tests to be made. After the custom of the time, a complete aircraft was

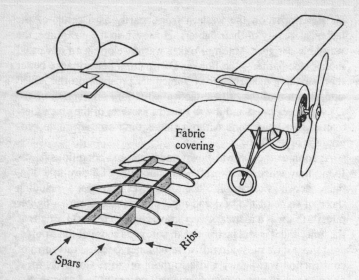

Figure 10. Fabric-covered monoplane wing.

mounted upside down in a test-frame and the wings were loaded
with piles of shot-bags, disposed so as to simulate the aerodynamic
loads which occur in flight. When tested in this way the wings
showed no sign of weakness, and they were broken only by a
load which was equivalent to six times the total loaded weight of
the aircraft. Although nowadays fighter aircraft are required to
withstand a load equivalent to twelve times their own weight, in
1917 a 'factor' of six was considered entirely adequate and almost
certainly represented a bigger load than would have occurred under the worst combat conditions at the time. In other words, the
aircraft should have been perfectly safe.

However, in the D8, when structural collapse did eventually
happen on the test-rig, the failure could be seen to begin in the
after of the two spars. To make quite certain, therefore, the
authorities ordered the rear spars of all Fokker D8s to be replaced by thicker and stronger ones. Unfortunately, after this had
been done, the accidents became more, not less, frequent, and so
the German Air Ministry had to face the fact that by 'strengthen-

ing' the wing by adding more structural material they had actually made it weaker.

By this time it was becoming clear to Antony Fokker that he was not going to get much effective help from the official mind. He therefore loaded up another D8 under his own supervision in his own factory. This time he took care to measure the deflections which occurred in the wing when it was loaded. What he found was not only that when the wing was loaded it deflected in bending (that is to say, the wing-tips would rise with respect to the fuselage when the plane was pulled out of a dive), but also that the wings twisted although no obvious twisting loads had been applied to them. What was particularly important was that the direction of this twisting was such that the aerodynamic incidence, or angle of attack of the wing, was significantly increased.

Pondering over these results that night, it suddenly occurred to Fokker that here lay the solution to the D8 accidents and to a great many other monoplane troubles as well. When the pilot pulled the control-stick back the nose of the plane rose and so did the load on the wings. *But* at the same time the wings twisted, so that air loads on the wings rose disproportionately; so the wings twisted more; so the loads rose still more; and so on, until the pilot no longer had any control over the situation and the wings were twisted off. Fokker had discovered something which is called a 'divergent condition' – which can also be a very lethal one.

What was actually happening in terms of elasticity?

Centres of flexure and centres of pressure

Consider a pair of similar, parallel, cantilever beams or wing-spars, joined together at intervals by horizontal fore and aft ribs bridging the gap between them (Figure 10). Suppose now a single upward force to be applied at some point on one of the outer ribs. Unless this force is applied at a point which is just half-way between the two cantilever spars (Figure 11), the load will not be equally shared between the spars and the upward force will be greater on one spar than on the other. If this happens then the more heavily loaded spar must deflect upwards further than its

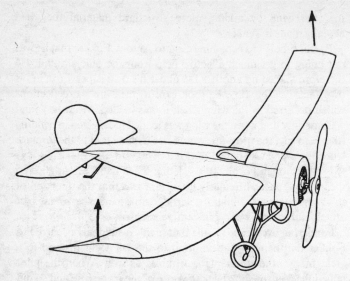

Figure 11. Coupled bending and torsion. *Only* if the vertical lift forces act
effectively at a point called the 'flexural centre' (in this case half-
way between the two spars) will the wings bend upwards without
twisting.

partner (Figure 12). In such a case the ribs joining the spars will
cease to be horizontal and the wing as a whole must twist. The
point at which a load must be applied so as to cause no twisting
in a beam-like structure is called the 'centre of flexure' or the
'flexural centre'.

Naturally, if there are more than two spars, or if the spars are of
differing stiffness, then the flexural centre will not be at the mid-
point but at some other position along the fore and aft or chord
line. *However, there is always a centre of flexure associated with
every sort of beam or beam-like structure.* A vertical load applied
at this point will not cause the beam or wing to twist; a load
applied at any other fore and aft position will cause a greater or
less amount of twisting or torsional deflection *as well* as the usual
bending deflection.

So far we have argued the case in terms of a single point load
applied to a beam or a wing. Naturally, the aerodynamic lifting

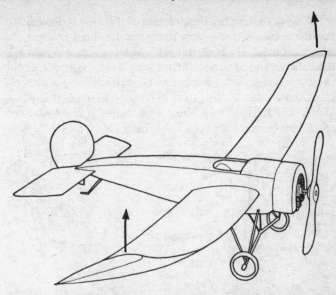

Fgure 12.   If the lift forces act at a point *away* from the flexural centre (e.g.
near the leading edge of a wing), then the wing (or any other
beam) will twist as it bends. If this causes an increase of aerodynamic incidence the result may be fatal, as it was in the
Fokker D8.

forces which, when an aircraft is in flight, press upwards on a wing
and so keep the machine in the air are diffused over the whole of
the wing surface. However, for the purposes of discussion and
calculation all these forces can be considered as acting together at
a single point on the wing surface which is known as the 'centre of
pressure' or C.P.

It might perhaps be supposed by the uninitiated that the C.P. of
the lift forces acting on a wing in flight lay at the middle of the
wing, half-way between the leading and trailing edges, that is to
say, at mid-chord. Actually it is a well-known fact of aerodynamic
life that this is just what does not happen. The centre of pressure
of the lift forces on a wing is really not far behind the leading
edge, usually near to what is called the 'quarter-chord' position:
that is to say, 25 per cent of the chord behind the leading edge.*

*This is why a dead leaf or a sheet of cardboard falls in the way it does.

It follows that, unless the structure of the wing is designed so that the flexural centre is close to the quarter-chord position, the wing must twist. The angle through which the wing will twist will naturally depend upon how stiff the wing is in torsion, but, on the whole, all wing-twisting is a bad and dangerous thing in an aeroplane and it is the designer's aim to reduce it as much as possible. This is why the quill of a bird's wing feather is usually located around the quarter-chord position (Figure 13).

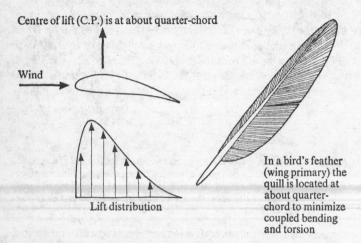

Centre of lift (C.P.) is at about quarter-chord

Wind

Lift distribution

In a bird's feather (wing primary) the quill is located at about quarter-chord to minimize coupled bending and torsion

Figure 13.   Lift distribution across an aerofoil.

In a simple fabric-covered monoplane wing both the position of the centre of flexure and also the torsional stiffness depend almost entirely upon the relative bending stiffnesses of the main spars. In the Fokker D8 the centre of flexure was a long way behind the centre of pressure and much too near mid-chord. The wing had not enough stiffness to resist the resulting torsional forces and so it was twisted off. Modifications which strengthened and stiffened the *rear* spar had the effect of moving the flexural centre still further backwards and so made the situation even worse. When these facts dawned on Antony Fokker he took the by now obvious step of *reducing* the thickness and stiffness of the rear spar, thus moving the centre of flexure further forward and closer to the C.P. When this was done the D8 became, comparatively speaking, a safe machine and a menace to the Royal Flying Corps and the French Air Force.

Because of the laws of aerodynamics the C.P. of the lift forces acting on an aeroplane wing must always be near to the quarter-chord position. To reduce the torsional or twisting stresses in the wing it is therefore necessary to design the structure in such a way that the centre of flexure is well forward in the wing and lies close to the C.P. However, the ailerons (which control the aircraft in roll, that is to say, when banking) apply large up or down forces to the wing tips, and these forces act at points not far from the trailing edge and thus a long way to the rear of the centre of flexure. Thus the ailerons inevitably exert large twisting loads on the wings every time the pilot banks the aircraft. It will be seen from Figure 14 that the direction of this twist is such as to change

Figure 14.   An aileron applies large vertical loads near the trailing edge of a wing and well *aft* of the wing's flexural centre. It therefore tends to twist the wing in such a way as to provide aerodynamic forces which are the *opposite* of those desired by the pilot.

the aerodynamic lift on the wing, as a whole, in the *opposite sense* to the action of the aileron and thus to reduce its effect. If the wing is not sufficiently stiff in torsion the effect of the aileron may actually be reversed, so that the pilot, wanting to roll or bank the aircraft to the *right*, and applying his controls in that sense, may find that the aircraft actually rolls to the *left*. This effect, which is not only disconcerting but also very dangerous, is called 'aileron reversal' and is not unknown. It is a serious problem in the design of modern fast aircraft. The cure or preventive is to ensure ample torsional stiffness in the wing structure.

In the early fabric-covered monoplanes, such as the D8, the torsional stiffness of the wings was almost entirely due to what is called the 'differential bending' of the two main spars. Not very much can be done about this and the amount of torsional stiffness which can be obtained from such a system – even with the help of a certain amount of wire rigging – is quite limited. For this reason such aircraft were always more or less dangerous – so much so that the authorities in nearly every country frowned on mono-plane construction, and in some cases it was actually forbidden.

The preference for biplanes was, therefore, not due to some kind of reactionary stupidity on the part of air ministries but rather to the fact that the biplane provides what is inherently a stiffer and stronger form of construction – especially in torsion. In practice, biplanes were both lighter and safer than monoplanes for many years, and in the early days the difference in speed was not very great.

What the strutted and braced biplane construction does is to provide, in effect, a sort of cage or 'torsion box' which is very strong and stiff, not only in bending but also in torsion. From Figure 15 it will be seen that the four main spars (two in each

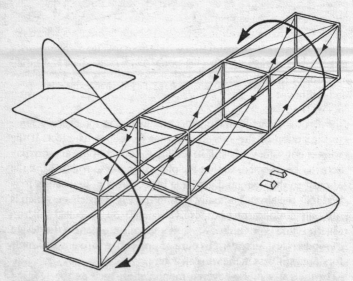

Figure 15.   Diagram of the main structure of a pair of wire-braced biplane wings subject to torsional forces, e.g. from the ailerons. The whole affair forms what is called a 'torsion box'.

wing) run along the corners of the box, while the spaces between them form a braced truss or lattice girder. One does not, of course, see the diagonal bracing on the top and bottom surfaces, because it is hidden by the fabric of the wings. However, this horizontal bracing is there all right, and its function is to take the shears which arise from the torsions in the wing structure. The manner in which such a box can resist torsion is shown diagrammatically in the figure. It will be seen that each side of the box is being sheared individually, very much like the lattice web of a trussed beam which is in bending. Notice that all four sides of the box are being sheared together and that they are mutually dependent. If one of the four sides were cut or removed there would be no resistance at all to torsion.

In a biplane these shear panels are necessarily made from struts and wires. However, if the structure did not have to fly but merely had to resist torsional forces on the ground, then the lattice of wires and struts could be replaced by continuous panels of metal or sheets of plywood. From a purely structural point of view the effect would be the same, just as it would be in the web of a beam truss. Torsion can therefore be resisted by any kind of box or tube whose sides may be continuous or alternatively of openwork lattice construction. In either case the walls or sides of the tube are subject to shearing stresses. In terms of weight and strength and stiffness this is a very much more effective way of resisting torsion than depending on the differential bending of two beams.

Formulae for the strength and stiffness in torsion of various kinds of rods and tubes are given in Appendix 3. Among other things it will be noticed that the strength and stiffness in twisting of a tube or torsion box depends upon the *square* of the area of its cross-section. Thus a torsion box of large cross-section, such as an old-fashioned biplane, will require little material and will be light in weight. When we build a modern monoplane, what we do is to turn the wing itself into a torsion tube with a continuous covering of metal sheet or plywood. However, even though we, perforce, use a much thicker wing than was the practice with biplanes, yet the cross-sectional area of the torsion tube, as a whole, is still much less than that of the biplane. So to get adequate torsional strength and stiffness we are forced to use comparatively thick and heavy skin. Thus a comparatively high proportion of the weight of the structure of modern aircraft has to be devoted to resisting torsion.

Although a lack of torsional stiffness is not quite as dangerous in cars as in aircraft, the character of a car's suspension and road-holding does largely depend upon it. The pre-war vintage cars were sometimes magnificent objects, but, like vintage aircraft, they suffered from having had more attention paid to the engine and the transmission than to the structure of the frame or chassis. These chassis, in fact, usually relied for any torsional stiffness which they might have had upon the differential bending of rather flexible beams – much like the old Fokker D8. It was the lack of stiffness in the chassis which gave these cars their highly uncertain road-holding characteristics and which made them so tiring to drive.

In an attempt to keep the wheels more or less in contact with the ground the springs and shock-absorbers of the vintage sports cars were stiffened up until they were virtually solid. As a result, of course, the ride became almost unbearably rough and jerky. Like the noisy exhaust, this kind of thing was no doubt impressive to the girl passenger, but it did not really do very much to keep the car on the road. The solution adopted by most modern car designers is to scrap the rather flimsy chassis and to take the torsion and bending loads through the pressed-steel 'saloon' body shell. This forms, with its roof, a big torsion box not wholly unlike the old biplanes. With so much stiffness at his disposal the designer can concentrate on providing a scientifically designed suspension which is both safe and comfortable.

As we have said, the strength and stiffness of a structure in torsion vary as the square of the area of its cross-section. This is more or less all right with bulky things like aircraft wings and ships' hulls and saloon cars; but when we come to shafts in engines and machinery the diameter – and therefore the area of the cross-section – is usually very limited, and so, as a rule, such members need to be made from solid steel. Even then, although they are often very massive, they are not always sufficiently strong. This is one of the reasons why engines and machinery are usually so heavy. As most experienced designers will tell you, any major requirement for torsional strength and stiffness in a structure is apt to be a curse and a blight. It puts up the weight and the expense and altogether provides a quite disproportionate amount of trouble and anxiety to the engineer.

Nature does not seem to mind taking a lot of time and trouble, and she has no sense at all of the value of money; but she is intensely sensitive to 'metabolic cost' – that is to say, to the price of a structure in terms of food and energy – and she is also generally pretty weight-conscious. It is not surprising, therefore, that she seems to avoid torsion like poison. In fact she nearly always manages to dodge out of any serious requirement for the provision of torsional strength or stiffness. As long as they are not subjected to 'unnatural' loads, most animals can afford to be weak in torsion. None of us likes having our arm twisted, and in normal life the torsional loads on our legs are small. However, when we attach long levers called skis to our feet and then proceed to ski rather badly, it is only too easy to apply large twisting forces to our legs. Because this is the commonest cause of broken legs in ski-ing, it has led to the development of the modern safety binding, which releases automatically in torsion.

Not only our legs, but virtually all bones, are surprisingly weak in torsion. Should you wish to kill a chicken – or any other bird – much the easiest way is to wring its neck. This is well known; what is less well known is how very weak are the vertebrae in torsion, as the beginner is apt to find out to his disgust and embarrassment when the head comes off in his hand. But then neck-wringing, like ski-ing, is an entirely artificial hazard and quite out of the ordinary course of nature. Unlike engineers, Nature has little interest in rotary motion and (like the Africans) she has never bothered to invent the wheel.

Chapter 13   The various ways of failing in compression

– or sandwiches, skulls and Dr Euler

By reason of the frailty of our nature we cannot always stand upright.

Collect for the 4th Sunday after Epiphany

As one would expect, the ways in which structures fail under compressive loads are rather different in their nature from the ways in which they break in tension. When we stress a solid in tension we are, of course, pulling its atoms and molecules further apart. As we do so, the interatomic bonds which hold the material together are stretched, but they can be safely stretched only to a limited extent. Beyond about 20 per cent tensile strain, all chemical bonds become weaker and will eventually come unstuck. Although the actual details of the tensile fracture process are complicated, it is broadly true to say that, when a sufficient number of interatomic bonds have been stretched beyond their breaking point, the material itself will break. The same sort of thing is also true when a material is broken by shearing. Strictly speaking, however, there is normally no analogous case of inter-atomic bond failure which is simply and directly due to compression. When a solid is compressed, its atoms and molecules are being pressed closer to each other, and under any ordinary conditions the repulsion between the atoms goes on increasing indefinitely as the compressive stress is raised. It is only when subjected to the enormous gravitational forces which exist in those stars which astronomers call 'dwarfs' that the compressive resistance between atoms collapses – with nightmarish consequences.*

* The result may be a concentration of mass so dense that its own gravitational field is strong enough to prevent, not only the escape of any matter' but also the departure of all forms of radiation. Thus no two-way communication is possible with such an area, and these regions of the Universe are for ever barred to us. These localities are known as 'Black Holes'. Like the island in Sir James Barrie's eerie play *Mary Rose*, they 'like to be visited'; but nothing can ever return.

Nevertheless, lots of very ordinary earthly structures do break by what is commonly described as 'compression'. What is really happening in failures of this sort is that the material or the structure finds some way of evading an unduly high compressive stress, usually by moving 'out from under' the load: that is to say, by running away in a sideways direction, using one of the escape routes which are practically always available. Looked at from the energy point of view, the structure 'wants' to get rid of an excess of compressive strain energy, and it will do so by means of whatever energy-exchange mechanism happens to be practicable in the circumstances.

Compression structures are thus apt to be rather shifty characters, and the study of compressive failure is more or less the study of ways of getting out of a tight place. As one might suppose, there are a number of different means of doing this. The escape method which the structure will use naturally depends upon its shape and proportions and upon the material from which it is made.

We have already discussed masonry at some length. Although buildings are essentially compression structures – and masonry must be kept in compression all the time – yet they cannot be said to fail by compression at all. Paradoxically, they can only fail by getting into tension. When this happens walls have a bad habit of developing hinge-points, as a consequence of which they tip up and fall down. Although arches are rather more stable and responsible structures than walls, they are capable at times of producing four hinge-points, after which they diminish both their strain energy and their potential energy by folding themselves up and reducing themselves to a heap of rubble. In any case, as we calculated in Chapter 9, the actual values of the compressive stresses in masonry are usually very low, far below the official 'crushing strength' of the material.

Crushing stresses – or the failure of short struts and columns in compression

However, if we take a brick or a block of concrete of fairly compact shape and subject it to a large compressive load – in a testing machine or by any other method – the material will eventually break in a manner which is conventionally called 'compression failure'. Although brittle solids like stone, brick, concrete and glass are generally crushed in such a way that they are reduced to fragments, or sometimes to powder, the failure is still not, in the strict sense, a compressive one. The actual fracture nearly always takes place by shearing. As we said in the last chapter, both tensile and compressive stresses necessarily give rise to shears at 45°; it is these diagonal shears which generally cause 'compressive failure' in short struts.

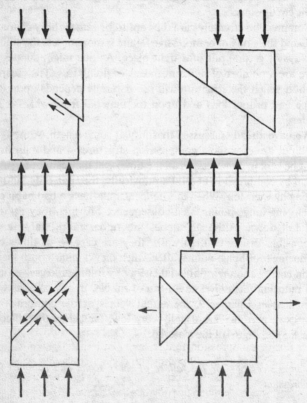

Figure 1. Typical 'compression failures' for a brittle solid such as cement or glass. Fracture is really due to shearing.

As we also said earlier on, all practical brittle solids are full of cracks and scratches and defects of one kind or another. Even if this is not the case when they are first made, such materials very soon become abraded from all sorts of virtually unavoidable causes. Naturally these cracks and scratches point in all directions in the material. It follows that a fair number of them will always be found to lie in directions which are diagonal to an applied compressive stress, that is to say, more or less parallel to the resulting shear stress (Figure 1).

Like tensile cracks, these shear cracks have a 'critical Griffith length'. In other words, a crack of a given length will propagate at a certain critical shear stress. When such conditions are reached in a brittle solid, such as concrete, the shear cracks will propagate suddenly, violently and perhaps explosively. When a shear crack has run diagonally across the width of a strut or other compression member, the two parts naturally slide past each other, so that the strut is no longer capable of carrying a compression load. The resulting collapse is likely to result in a large release of energy, and this is why brittle materials like glass and stone and concrete throw out splinters, which can be dangerous, when they are crushed or hit with a hammer. In fact the release of strain energy is quite often large enough to 'pay' for reducing the material to a powder. This is what happens when we crush lumps of sugar with a hammer or a rolling-pin.

The failure of a ductile metal – or, come to that, of butter or plasticine – under compressive stress is due to similar causes. What happens is that the metal 'slips' or slides (because of the dislocation mechanism) within itself under the shearing stress. Again this happens along planes roughly at 45° to the compressive load: thus a short metal strut bulges outwards into some barrel-like shape (Figure 2). Because of the high work of fracture of ductile metals, such materials are far less likely to throw off splinters during compression failure, and the immediate consequences of the fracture are likely to be less dramatic and a good deal less dangerous. It is this effect, the tendency to bulge under compression, which we make use of when we spread the head of a metal rivet by hammering it or by squeezing it in a hydraulic press.

Materials like wood and the artificial fibrous composites such

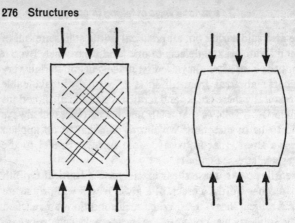

Figure 2. Failure of a ductile material, such as a metal in compression. Failure is again due to shearing, but this time the effect is to cause the metal to bulge.

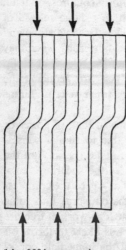 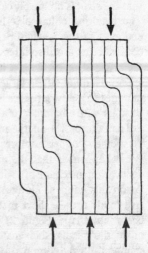

(a)   90° 'compression crease'   (b)   Diagonal 'compression crease'

Figure 3. Failure of a fibrous material such as wood or Fibreglass in compression. Note that the 90° crease involves a volume contraction and can therefore only take place in a material containing voids, such as wood. 'Solid' composites must fail by mode (b), which does not involve a change of volume.

as Fibreglass and carbon fibre materials generally fail in compression in a rather different way. In such cases the reinforcing fibres 'buckle' or fold in sympathy with each other under the compressive load, so that what is called a 'compression crease' runs across the material. These compression creases may run either diagonally or at 90° to the direction of the applied compressive stress or sometimes at various angles in between (Figure 3). Unfortunately compression creases often tend to form in fibrous materials at quite low stresses. These materials are therefore sometimes 'weak in compression', and this point needs to be considered when using them.

Breaking stresses of materials in tension and in compression

The various text-books and reference books generally make a great parade of tabulating the 'tensile strengths' of common engineering materials. As a rule, however, these books are a good deal more reticent about compressive strengths. This is partly because the experimental values of the compressive failing stresses of materials vary much more with the shape of the test-piece which has been employed than do the tensile strengths. Sometimes this effect is so great that it becomes almost meaningless to quote a figure. However, although a cautious attitude to compression strengths is in some ways justified, it does have the effect of glossing over some of the facts of structural life. One of these facts is that there is really no consistent relationship at all between the tensile and the compressive strength of a material.* Some rather approximate figures for common materials are given in Table 5. The compressive strength values are those which might be obtained using test-pieces having a ratio of length to thickness of something like three or four to one. For specimens much fatter or thinner than this the breaking stresses might be quite different.

---

*In so far as failure in both tension and compression tend to occur by shearing – as in ductile metals – the tensile and compressive strengths would be identical. However, there are so many exceptions to this rule as to make it practically valueless.

One of the obvious lessons to be drawn from Table 5 is that, when we come to design a thing like a beam which is stressed in

TABLE 5

Some materials with unequal tensile and compression strengths. (These figures are approximate.)

Material	Tensile strength p.s.i.	MN/m²	Compressive strength p.s.i.	MN/m²
Wood	15,000	100	4,000	27
Cast iron	6,000	40	50,000	340
Cast aluminium	6,000	40	40,000	270
Zinc die castings	5,000	35	40,000	270
Bakelite, polystyrene and other brittle plastics	2,000	15	8,000	55
Concrete	600	4	6,000	40

both tension and compression, we may need to watch our step. It may be necessary to design a beam which has a highly asymmetrical section. In Victorian cast-iron beams the tension side is

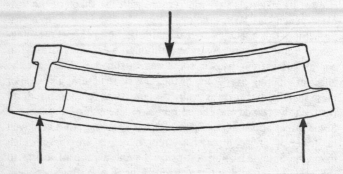

Figure 4. Cast-iron beams are usually made thicker on the tension face than on the compression face because cast iron is weaker in tension.

usually very much thicker than the compression side – because cast iron is weaker in tension than in compression (Figure 4). Contrariwise, the wing-spar of a wooden aircraft, such as a sailplane, is always much thicker on the upper or compression side, since wood is weaker in compression than in tension (Figure 5).

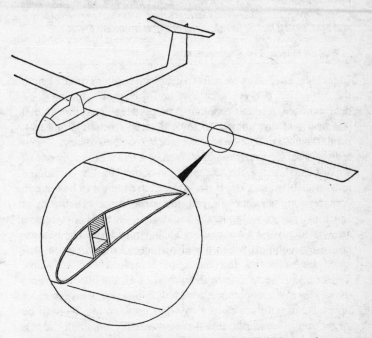

Figure 5.   A wooden glider wing-spar is usually made thicker on the compression side than on the tension face because wood is weak in compression.

The compressive strength of timber and of composite materials

He said he had been making masts for over fifty years, and, as far as he knew, they had all been sound spars. He said I was the only man he had ever met who deliberately planned to ruin a good mast by cutting the heart right out in a most sensitive spot. He said that any man who could do a thing like that would – (and here I tone down his words a lot) – curse aloud in church, wipe his nose on the table-cloth, take soundings in a cess-pit and eat the arming.

. . . And that was that. Both George and I thought secretly that the spar

was a great deal too whippy for comfort, but in the face of those experts we decided it might be wise to keep our opinions locked up within us. Which was well. For the experts were expert. Later on, when our main shrouds did carry away in a wicked Gulf Stream squall, that mast bent – and bent – and bent, until it looked like the letter S; but it would not break.

Weston Martyr, *The Southseaman*

In real life, as soon as we start to deal with columns of any length the distinction between a column and a beam becomes a good deal confused. A longish column – such as a leg-bone of an animal – is nearly always subject to some degree of bending, and as a result the material on the concave side is compressed more than it is elsewhere. Contrariwise, in a beam or a truss, especially one of sophisticated design, the 'compression boom' must be considered as a strut. In either case, if the material itself tends to be weak in compression, whether we call the structure a 'beam' or a 'column', failure will generally begin when the total compressive stress at the worst place reaches a dangerous level. The best examples of columns which are also subject to bending are provided by trees and the masts of traditional sailing ships. Tree-trunks have to sustain the weight of all the bits and pieces of the tree in direct compression, but, in practice, the stresses set up by bending forces caused by wind pressure are likely to be larger and more important. Again, masts are nominally struts, carrying only axial compression, but, because of the stretching of the rigging, and for other causes, they are in fact subject to a good deal of bending, especially if anything in the rigging should happen to break.

The masts of big ships like H.M.S. *Victory* had to be built up by joining many pieces of wood together with iron hoops, but for masts of more moderate size the traditional spar-makers preferred to use single pine or spruce trees, left as nearly as possible in their original condition. Not only did these craftsmen strongly resist any suggestion that a mast should be built up or hollowed out in such a way as to produce a more 'efficient' tubular section; they also took care to remove as little as possible – beyond the bark – of the outer surface of the tree. In other words they tried, as far as they could, to use the tree in its natural state.

For many years professional engineers, who knew all about beam theory and neutral axes and second moments of area, despised this as so much traditional nonsense. In fact the first thing that a modern engineer does with a tree is to cut it up into

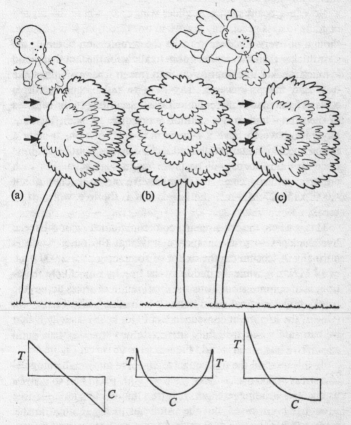

Figure 6. (a) Tree bent by the wind with *no* pre-stress in wood. Stress distribution across the trunk is linear and maximum tension and compression stresses are equal.

(b) Pre-stressed tree in a calm. The outside of the trunk is in tension all round; the inside is in compression.

(c) Pre-stressed tree in a strong wind. Compression stress is *halved* and this tree can bend *twice* as far as the one in (a).

small pieces, which he then glues together again – preferably into some kind of hollow section. It is only recently that we have realized that, after all, the tree does know a thing or two. Among other subtleties, the wood in various parts of the trunk grows in such a way that it is 'pre-stressed'.

Now in a beam such as a glider wing-spar, where the biggest bending load is practically always in one direction, it is possible, though not very efficient, to make the compression boom of the spar thicker than the tension boom to allow for the fact that wood is much weaker in compression than it is in tension. Things like trees and masts, however, may have to resist bending forces coming from many different directions – according to the caprices of the wind – and so this solution is not open to them. Trees, at any rate, have to have a symmetrical cross-section, usually a round one. For an un-prestressed section the distribution of stress under bending loads will be linear, as in Figure 6a. For such an arrangement, when the compressive stress reaches about 4,000 p.s.i. (27 MN/m^2) the beam, that is the tree, will start to break.

This is where the pre-stressing comes in. Somehow or other the tree manages to grow in such a way that the outer wood is normally in tension (to the extent of something over 2,000 p.s.i. or 14 MN/m^2), while the middle of the tree, by way of compensation, is in compression. Thus the distribution of stress across the trunk, under normal conditions, is something like Figure 6b. (One of the important consequences of Hookean elasticity is that we can safely and truthfully superpose one stress system upon another.) Thus, when we add Figures 6a to 6b we get Figure 6c.

By this method the tree roughly *halves* the maximum compressive stress (4,000 p.s.i. − 2,000 p.s.i. = 2,000 p.s.i.) and so *doubles* its effective bending strength. It is true that the maximum tensile stress has been raised, but the wood has plenty in hand in this respect. What the tree does in the way of protecting itself by pre-stressing is exactly the opposite of what we do when we make a pre-stressed concrete beam. In the latter case the concrete is weak in tension and relatively strong in compression; the danger is that, when the beam is bent, failure may occur in the concrete on the tension face. To avoid this we put the steel reinforcing rods, which are inside the beam, permanently into tension, so that the concrete is permanently in compression. Thus the beam has to be bent considerably before the compressive stress in the concrete near the surface is relieved and replaced by a tension stress. Thus the cracking of the cement is postponed, since the beam has to be bent further before the critical tensile strain is reached.*

As we have said, both timber and the fibrous composite materials generally fail in compression by the formation of bands or creases of bent and buckled fibres. My colleague Dr Richard Chaplin points out that these compression creases have a good deal in common with cracks which occur in tension. In particular they are often started by stress-concentrations at holes or other defects in the material. In general, fastenings like nails and screws do not much weaken timber, always provided that they are in place and fit tightly. Once they are removed, however, the resulting hole has a much more serious effect; and no doubt the same is true of knots in timber. In a highly stressed wooden structure, such as a glider or a yacht's mast, it is therefore wise to leave unwanted nails and screws alone and not try to pull them out. If needs be, they can be cut off flush with the surface of the wood.

Furthermore, as Richard Chaplin says, the formation of compression creases in a fibrous material requires energy. In fact the amount of energy required is rather larger than the work of fracture of the material in tension. It follows that the propagation of compression creases needs a supply of strain energy and that their behaviour is something like that of a Griffith crack. There are, however, some important differences.

We have said that, in materials of the kind we have been discussing, compression creases can occur both at 45° and also at 90° to the direction of loading. (They can also occur at other angles between 45° and 90°.) The 45° crease is effectively a shear crack, and, if the conditions are right, it will spread right across the material, much like a Griffith crack in shear. However, the 90°

---

*Note that many seaweeds, which are made largely from alginic acid – a weak and brittle substance – are pre-stressed in the same sense as reinforced concrete. Just as reinforced concrete economizes in steel, so seaweeds economize in the scarce, strong component, cellulose.

crease is shorter – and therefore consumes less energy – for a given depth of penetration below the surface of the material.

For this reason the 90° crease is, on the whole, more likely to occur. However, although the 90° crease seems to be easier to start off, it is more likely to come to a halt after travelling for a short distance. This is because, as the crease advances, its two sides tend to get pinched together (or 'come up solid') and so cease to release much strain energy. Thus complete failure is unlikely to take place, at any rate immediately.

What may happen in these circumstances is that many little creases will form, one behind the other, all along the compression surface of a beam. This can be seen on the compression face of a wooden bow, and sometimes with oars (Figure 7). Although

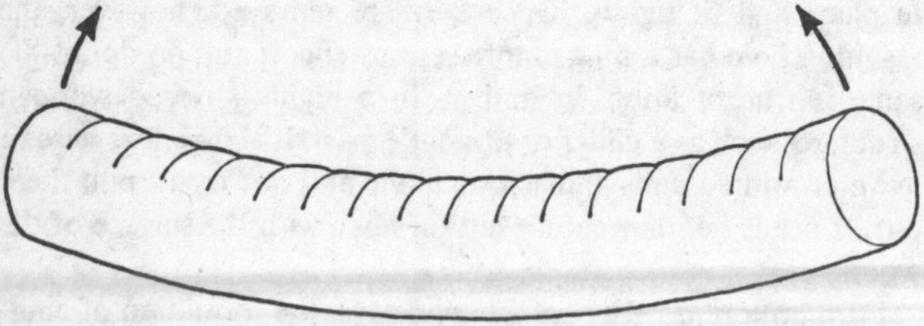

Figure 7. Multiple compression creases on the compression face of a round piece of timber such as a tree, a mast, an oar or a bow. These creases may not be able to spread, and so complete fracture does not occur.

engineers often advocate 'efficient' H sections or box sections for beams, this can be a mistake. For reasons which are easily demonstrated,* the strain energy release conditions are often less favourable to the propagation of both cracks or compression creases when the beam section is rounded – like a tree – and this is probably the rationale behind the rounded cross-sections of most wooden bows. No doubt something of the kind is also relevant to the rounded cross-sections of the bones of animals.

* As a crack or a compression crease with a straight front (like a saw cut) penetrates across a round section its surface area may increase more rapidly than the rate of release of strain energy from the material behind it; and so Griffith is frustrated.

So long as the material is stressed consistently in compression there are many hindrances to the spread of compression creases. This is one of the reasons why wood is generally such a safe material. However, under conditions of reversed loading, it can be very dangerous indeed. This is because the buckled fibres which constitute a compression crease have little or no tensile strength, and so, under tension forces, the crease acts like an ordinary crack. It is especially dangerous because, in tension, there is now no restriction on the release of strain energy since the two sides of the crack are free to spring apart.

One of the best ways to arrange for a wing to come off a wooden glider in flight is to make a heavy landing with it. If one puts the aircraft down with a really bad bump, the wings will, momentarily, be bent downwards towards the ground. This may cause compression creases in the wood of what is normally the tension part of the main spar. If this happens, the creases are most unlikely to be spotted during routine inspections. The next time the glider is flown the spar may break in tension at this point, after which, of course, the wing will fall off.

Leonhard Euler and the buckling of thin struts and panels

What we have said so far applies to struts and other compression members which are fairly short and thick. As we have seen, these usually fail in compression by a diagonal shearing mechanism, or sometimes by the formation of local creases in the fibres. However, a large number of compression structures of one sort or another involve members which are long and thin and which fail in a totally different way. A long rod, or a membrane such as a thin sheet of metal or a page of this book, fails in compression by buckling, as can very easily be seen by doing the simplest experiment. (Take a sheet of paper and try to compress it lengthwise.) This mode of failure – which has important technical and economic consequences – is called 'Euler* buckling' since it was originally analysed by Leonhard Euler (1707–83).

Euler came from a German-Swiss family well-known for its mathematical ability, and he very soon acquired fame as a

* Pronounced 'Oiler'.

mathematician: so much so that, while still quite young, he was invited to Russia by the Empress Elizabeth. He spent most of his life at the Court of St Petersburg, taking refuge for a time with Frederick the Great at Potsdam when the political situation in Russia got too exciting. Life at the courts of the Enlightened Despots in the middle of the eighteenth century must have been both interesting and colourful, but little of this is reflected in Euler's voluminous writings. As far as I can trace, there appears to be no incident of any noticeable human interest recorded of him in any of his biographies.* He simply went on for a very long time doing mathematics and writing it all down in an enormous number of learned papers, the last of which were still being published forty years after his death.

As a matter of fact, Euler did not really mean to do anything about columns at all. What happened was that, among a great many other mathematical discoveries, he had invented something called the 'calculus of variations', and he was looking for a problem to try it out on. A friend suggested that he might use this method to calculate the height of a thin vertical pole which would just buckle under its own weight. It was necessary to make use of the calculus of variations to tackle this rather hypothetical problem because, as we mentioned in Chapter 3, the concepts of stress and strain were not invented until much later.

Put in modern terms, what Euler came up with was what we now call 'the Euler formula for the buckling load of a strut', which is

$$P = \pi^2 \frac{EI}{L^2} \text{ (See Figure 9.)}$$

where P = load at which the column or panel will buckle

E = Young's modulus of the material

I = second moment of area (the so-called 'moment of inertia') of the cross-section of the strut or panel (Chapter 11)

L = length of strut.

Naturally, all these quantities must be in mutually consistent units.

---

* Except, of course, his increasing blindness in later life.

(It is curious, but convenient, that so many of these important structural formulae should be, algebraically, so very simple.*)

Euler's formula applies to all sorts and kinds of long, thin columns and struts – both solid and hollow – and, perhaps even more importantly, to thin panels and plates and membranes such as occur in aircraft and ships and motor cars.

Thus, if we plot the failing load of a strut or a panel against its length we get a diagram something like Figure 8, which shows two

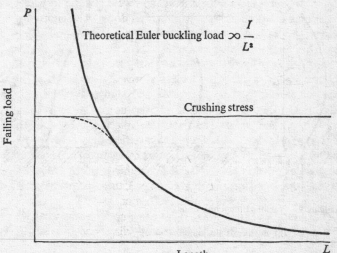

Figure 8.   Variation in the compressive strength of a column with its length.

regimes of failure. For a short strut, failure will be by crushing. When the ratio of length to thickness increases to a value between about five and ten, then this line will be crossed by the curve which represents Euler buckling failure. Buckling now becomes the weaker mode, and so long struts will fail in this way. In practice the change-over from crushing failure to Euler buckling is not a

---

*Several modern proofs of Euler's formula are to be found in the textbooks. See, for instance, *The Mechanical Properties of Matter* by Sir Alan Cottrell.

sharp one and there will be a transitional region, something like the dotted line in the diagram.

The form of the Euler formula which has just been given assumes that the strut or panel is 'pin-jointed', or free to hinge, at both ends (Figure 9). Usually, anything which tends to prevent a

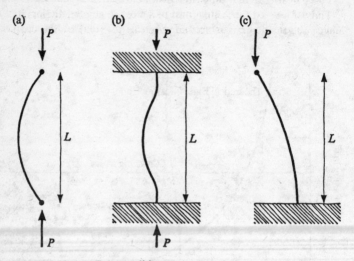

Figure 9.   Various Euler conditions.

    (a) Both ends pin-jointed. $P = \pi^2 \dfrac{EI}{L^2}$

    (b) Both end fixed in direction and position. $P = 4\pi^2 \dfrac{EI}{L^2}$

    (c) One end *encastré*, the other pin-jointed and free to move sideways. $P = \pi^2 \dfrac{EI}{4L^2}$

strut or panel from hinging at the ends will increase the buckling load. For the extreme case, where both ends are rigidly restrained, the buckling load, P, is multiplied by as much as 4. Very frequently, however, the achievement of any considerable degree of end restraint involves extra weight and complication and cost and may not be worth doing. Furthermore 'rigid' end-connections will transmit any misalignment of the end-attachments to the strut. If this happens, the strut may be prematurely bent, and so, in practice, made weaker. For this reason the 'rigid' stepping of masts, by attaching them both to the deck and the keel, is no longer usual (Figure 10).

It will be noticed that, in the Euler formula which we have just written down, there is no term which represents a breaking stress. The buckling load of a strut or a panel of a given length depends solely upon the 'I' (or second moment of area) of the cross-

Figure 10.   If a column is clamped at the ends in such a way as to force it out of alignment, its buckling load may be reduced. Since rigging is liable to stretch, it is no longer customary to fix masts at both deck and keel.

section and upon the Young's modulus or stiffness of its material. A long strut does not 'break' when it buckles. It just bends elastically in such a manner as to get out of the way of the load. If the 'elastic limit' of the material has not been exceeded during buckling, then, when the load is removed, the strut will simply spring straight again and recover its original shape, quite undismayed by its experience. This characteristic can often be a good thing, for it is possible to design 'unbreakable' structures in this way. Broadly speaking, this is how carpets and doormats work. Predictably, Nature uses the principle very widely, especially for small plants like grasses which inevitably get trodden on. This is why it is possible to walk on a lawn without doing it any harm. It is the ingenious combination of spiky thorns with Dr Euler's principle which makes a quickset hedge practically indestructible and impenetrable to both men and cattle. On the other hand, mosquitoes and other insects which make use of long slender stabbing weapons have to employ an indecent amount of low structural cunning to prevent these thin struts from buckling when they sting you.

During Euler's lifetime the actual technological uses for his formula were very few. Practically the only important application would have been in the design of ships' masts and other spars. However, contemporary shipwrights had already got this problem under control in a pragmatic way. The magnificent eighteenth-century text-books on shipbuilding, such as Steele's *Elements of Mastmaking, Sailmaking and Rigging*, contain extensive tables of the dimensions of every kind of spar, based on experience, and it is doubtful if these recommendations could have been much improved upon by calculation.

Serious interest in buckling phenomena only began about a century after Euler's time and was largely due to the increasing use of wrought-iron plates in constructional work. These plates were naturally much thinner than the masonry and woodwork to which engineers had been accustomed. The problem was first tackled seriously in the case of the Menai railway bridge, about 1848. The design of this bridge was the joint responsibility of three great men, Robert Stephenson (1803–59), Eaton Hodgkinson (1789–1861), a mathematician and one of the first professors of engineering, and Sir William Fairbairn (1789–1874), a pioneer in the structural use of wrought-iron plates.

Stephenson's railway suspension bridges had been a failure because they were too flexible. Furthermore, the Admiralty, not unreasonably, insisted upon a clear 100 feet (30 metres) headroom beneath the bridge for shipping. The only way of combining the necessary stiffness with the headroom which was demanded seemed to be to design a beam bridge far longer than had ever been built before. For various reasons it seemed best to make the beams, each of which had to be 460 feet (140 metres) long, in the form of tubes built up from wrought-iron plates, with the trains running inside the tubes.

It fairly soon became evident that one of the most serious design problems lay in the buckling of the iron plates which formed the upper or compression side of the beams. Although Euler's formula is accurate enough for simple panels and struts, the shape of the bridge tubes was necessarily complicated, and no adequate mathematical theory existed at that time. The three designers had thus no option but to experiment with models. As might have been expected, these proved to be confusing and unreliable – so much so that the three men quarrelled among themselves and at one time it looked as if the partnership would break up with no

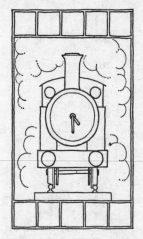

Figure 11. Britannia bridge: tubular box beam.

really safe design for the tubes in sight. Eventually, however, a cellular box beam was decided upon (Figure 11). To everybody's immense relief this proved satisfactory; it lasted until 1970.

Since Stephenson's time, a very great amount of mathematical research has been carried out on the buckling of thin shells; but the design of such structures is still accompanied by even more than the usual degree of uncertainty. So the development of critical structures of this kind is likely to be expensive, because of the full-scale strength tests which may be needed before the design can be finalized.

Tubes, ships and bamboos – and something about Brazier buckling

Since, according to Euler, the buckling load of a strut varies as EI/L^2, the compressive strength of a long column is liable to be very low indeed. The only thing we can do about this is to increase EI – if possible in proportion to L^2. For most materials E, the Young's modulus of elasticity, is pretty well constant, so what we have to do in practice is to increase I, the second moment of area of the cross-section. This means that we have got to make the

Figure 12. 'Brazier' or local buckling of a thin-walled tube under axial compression.

column fatter. That, of course, is exactly what we do in masonry, for instance in the sturdy columns of a Doric temple. The result, however, is excessively heavy, and if we want to make a light structure then we shall have to design some sort of expanded section. This sometimes takes the form of an 'H' or star shape, or sometimes a square box. On the whole, however, round tubes are usually better and more efficient.

The use of tubes is extremely popular both with engineers and with Nature, and tubular struts are very widely used for all sorts of purposes. However, a tube under compression has a choice of *two* modes of buckling. It may buckle in the way we have been describing: that is to say, in a long-wave mode, over its whole length, Euler-fashion. Alternatively, it may buckle in a short-wave mode, that is to say, locally, by putting a sort of crease or crumple into the wall of the tube. If the radius of the tube is large and if the wall is thin, then the strut may well be safe against Euler, or long-wave, buckling; but it will fail by the local crumpling of the skin. This is easily demonstrated with a thin-walled paper tube. One form of this local buckling or crumpling is called 'Brazier buckling' (Figure 12). It is this effect which sets a limit to the use of simple tubes and thin-walled cylinders in compression.*

The commonest way to guard against Brazier buckling is to stiffen the skin of a thin-walled structure by attaching extra members, such as ribs or stringers, to it. Stiffeners which run circumferentially are generally called 'ribs', while those which run lengthwise are called 'stringers' (except by botanists, who will call them 'ribs'). The shell-plating of ships is traditionally stiffened by means of ribs and bulkheads, though, recently, large tankers have been built on the 'Isherwood' system, which largely depends upon longitudinal stringers. A sophisticated shell structure, such as an aircraft fuselage, is usually stiffened by both ribs and stringers. The hollow stems of grasses and bamboos, which tend to flatten when they are bent, are very elegantly stiffened by means of 'nodes' or partitions or bulkheads, spaced at intervals along the stem (Figures 13 and 14).

---

*In a thin-walled circular tube local buckling will generally occur when the stress in the skin reaches a value equivalent to

$$\frac{1}{4}E\frac{t}{r}$$

where t = wall thickness
r = radius of tube
E = Young's modulus.

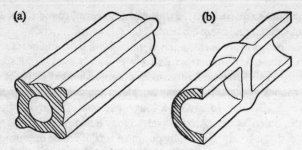

Figure 13. Two ways of stiffening a hollow plant stem against local buckling.
(a) Longitudinal stringers.
(b) Nodes or bulkheads – common in grasses and bamboos.

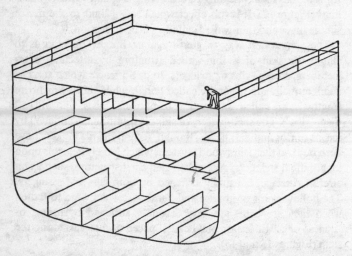

Figure 14. Engineering shell structures such as ships and aircraft generally
use both stringers and ribs or bulkheads. This is a diagram of
the Isherwood construction often used in oil tankers.

Leaves, sandwiches and honeycombs

Thin plates and panels and shells are continually cropping up both
in Nature and in technology, and, the larger and the thinner these

structures are, the more likely they are to deflect or crumple under bending and compressive loads. In principle, anything which stiffens a column or a panel in bending will also increase its resistance to buckling and so make it stronger in compression. One way of doing this is by staying a strut or a panel with ropes or wires; this is a solution which is never used in plants. Alternatively, and perhaps preferably, one can stiffen the member with ribs or stringers, by corrugating it, or by making it of cellular construction.

Wood is a cellular material, and so are most other plant tissues, notably the stem-walls of grasses and bamboos. Furthermore, in the competitive struggle for existence, many plants depend critically upon the structural efficiency of their leaves, because they must try to expose the maximum area to sunlight, for photosynthesis, at the minimum metabolic cost. Leaves are therefore important panel structures, and they seem to make use of most of the known structural devices to increase their stiffness in bending. Nearly all leaves are provided with an elaborate rib structure*; the membranes between the ribs are stiffened by being of cellular construction, and in some cases they are further stiffened by corrugations. In addition to all this, the leaf as a whole is stiffened hydrostatically by the osmotic pressure of the sap.

In engineering structures, panels and shells are very often stiffened by means of ribs or stringers which are glued or riveted or welded to the plating, though this is not always the lightest or the cheapest way of doing the job. Another way of tackling the problem is to make the shell-plating in two separate layers which are then spaced apart by being glued to some kind of continuous support, usually made as light as possible. Arrangements of this kind are called 'sandwich constructions'.

In modern times sandwich panels were first used for serious constructional purposes by Mr Edward Bishop, de Havilland's famous chief designer, for the fuselage of the now-forgotten Comet aircraft of the 1930s.† It is probably best known for its use in the

---

*The ribs of the leaf of the Victoria Regia lily are traditionally supposed to have inspired Sir Joseph Paxton's design for the Crystal Palace in 1851.

†Which had no direct connection with the later jet airliner of the same name.

successor to this aeroplane, the war-time Mosquito. In both these aircraft the core of the sandwich was made of light-weight balsa-wood, with skins of heavier and stronger birch plywood glued to either side.

Though the Mosquito was a most successful aircraft, balsa-wood is apt to soak up water and rot; moreover, supplies of this rather soft and fragile tropical wood are limited in quantity and variable in quality. As things turned out, research on core materials for sandwich shells and panels was much stimulated at about this time by another factor altogether; this was the intro-duction of airborne radar. With this equipment the moving radar reflector or 'scanner' had to be housed and protected by putting it inside a large streamlined dome or fairing, which soon came to be known as a 'radome'. Naturally these fairings had to be transparent to high-frequency radio waves, and this meant that, in practice, they had to be made from some sort of plastic, usually Fibreglass or Perspex. The transparency of the radome shell to radar could be much improved – at least in theory – by the use of a sandwich construction whose thickness was carefully related to the wavelength of the radiation which was being transmitted – in exactly the same way as the thickness of the coating or 'blooming' on a modern camera lens is related to the wavelength of visible light.

Damp balsa, like any other damp wood, is nearly opaque to radar; and under war-time conditions balsa is practically always

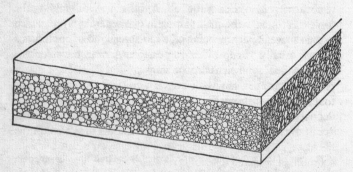

Figure 15.   Foamed resins are often used as light-weight core materials in sandwich constructions.

damp. This ruled out its use for radomes, and so it was necessary to develop more waterproof light-weight materials. This was done by 'foaming' artificial resins of various kinds. The result looked something like a meringue or 'Aero' chocolate (Figure 15). A good many foamed resins of this kind were developed; they have a number of virtues, and they were used not only for the cores of radome sandwiches but for all sorts of other structural sandwich panels as well. Some of them are still in use today. They are used, for instance, in boatbuilding because the walls of their cells or cavities are nearly impervious to water. However, for the cores of sandwich panels of the highest structural efficiency, resin foams are rather heavier and rather less stiff than one might wish. In other words, the market for light-weight core materials was more or less an open one.

One day, towards the end of 1943, a circus proprietor called George May called to see me at Farnborough. After he had told me several Gerald Durrell-type stories about the difficulties of keeping monkeys in travelling circuses, he produced something which looked like a cross between a book and a concertina. When he pulled on the ends of this invention, the whole thing opened out like one of those coloured-paper festoons which people use for Christmas decorations. It was in fact a sort of paper honeycomb of very light weight but of quite surprising strength and stiffness. Did I think that such a thing could be of any use in aircraft? The snag, as George May modestly admitted, was that, since it was only made from brown paper and ordinary gum, it had no moisture resistance at all and would fall to bits if it got wet.

This must have been one of the relatively few occasions in history when a group of aircraft engineers have been seriously tempted to throw their collective arms around the neck of a circus proprietor and kiss him. However, we resisted the temptation and told May that there could be no serious difficulty in waterproofing the paper honeycomb by means of a synthetic resin.

This was exactly what we did (Figure 16). The paper from which the honeycomb was to be made was impregnated before use with a solution of uncured phenolic resin. After the honeycomb had been made and expanded, the resin was cured and hardened by baking it in an oven. As a result the paper was not only made

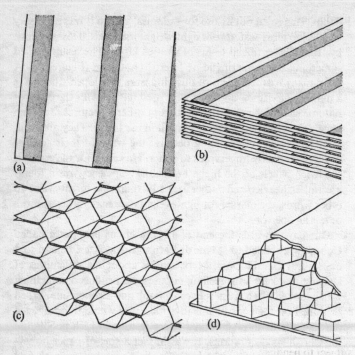

Figure 16.   Construction and use of paper honeycomb.
(a) Resin-impregnated paper is printed with parallel stripes of glue.
(b) Many sheets are glued together into a thick block with glue stripes staggered.
(c) When the glue is set, the block of material is expanded into a honeycomb. After this the resin is hardened.
(d) Slabs of honeycomb are glued between sheets of ply, plastic or metal to form a structural sandwich.

waterproof but also strengthened and stiffened. This material was very successful and was used in the cores of sandwiches for all kinds of military purposes. Though it is not used a great deal in aircraft nowadays, something like half the household doors in the world are made by gluing thin sheets of plywood or plastic on either side of a paper honeycomb. It is even more widely used abroad, especially in America, than in England, and the world

production of paper honeycomb must be very considerable.
   Although the use of sandwich construction, foamed resin cores
and honeycombs is relatively new in engineering, it has been used

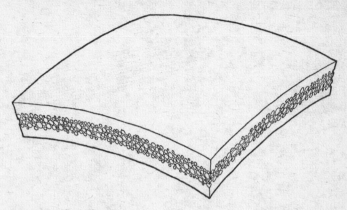

Figure 17.   Cancellous bone.

for a very long time in biology. What is called 'cancellous' bone
(Figure 17) exploits this principle. Each of us carries around quite
a good example in the bones of our skulls, which are, of course,
subject to bending and buckling loads.

Part Four

And the consequence was . . .

Chapter 14   The philosophy of design

– or the shape, the weight and the cost

Philosophy is nothing but discretion.

John Selden (1584–1654)

As we have seen, very much the commonest day-to-day practical use of structural theory is in analysing the behaviour of some specific structure: either one which it is proposed to build, one which is actually in existence but whose safety is in question, or else one which has, rather embarrassingly, already collapsed. In other words, if we know the dimensions of a given structure and the properties of the materials from which it is made, we can at least try to predict how strong it ought to be and how much it will deflect. However, although calculations of this sort are clearly very useful in particular instances, this kind of approach is only of limited help to us when we want to understand why things are the shape they are or when we want to choose which, out of several different classes of structure, would be best for a particular service. For instance, in making an aeroplane or a bridge, would it be better to use a continuous shell structure made from plates or panels or else a criss-cross lattice arrangement built up from rods or tubes and braced, perhaps, with wires? Again, why do we have so many muscles and tendons and comparatively few bones? Furthermore, how is the engineer ever to select from the large variety of materials which are usually available? Should he make his structure from steel or aluminium, from plastic or from wood?

The 'design' of plants and animals and of the traditional artefacts did not just happen. As a rule both the shape and the materials of any structure which has evolved over a long period of time in a competitive world represent an optimization with regard to the loads which it has to carry and to the financial or the metabolic cost. We should like to achieve this sort of optimization in modern technology; but we are not always very good at it.

It is not widely realized that this subject, which is sometimes called the 'philosophy of design', can be studied in a scientific way. This is a pity, because the results are important, both in biology and in engineering. Although not much regarded, the study of the philosophy of design has, in fact, been going on for quite a number of years. The first serious engineering approach to the subject was made by A. G. M. Michell around 1900.* Though biologists had been making remarks about the 'square-cube law' (Chapter 9) practically since it was propounded by Galileo, it was not until 1917 that Sir D'Arcy Thompson published his beautiful book *On Growth and Form* (still in print), which was the first general account of the influence of structural requirements on the shapes of plants and animals. For all its many virtues, the book is not a very numerate one, and the engineering views expressed are not always sound. Though greatly, and justly, praised, *Growth and Form* did not have much real influence on biological thinking, either in its own time or for long afterwards. It does not seem to have influenced engineers very much either, no doubt because the time for an interaction between biological and engineering thought was not ripe.

In recent years the chief exponent of the mathematical study of the philosophy of structures has been H. L. Cox. Besides being a distinguished elastician, Mr Cox has the additional merit of being an expert on Beatrix Potter. I hope that he will forgive me for saying that he is in some ways a little like the great Thomas Young. For he shares not only something of Young's genius, but also a good deal of Young's obscurity of presentation. I am afraid lesser mortals often find Cox's expositions difficult to follow without the aid of an evangelist or interpreter. This may account for the fact that his work has received less attention than it deserves. Much of what follows is based on Cox, directly or indirectly. Let us begin with his analysis of tension structures.

The design of tension structures

It is a curiosity of engineering design that it is impossible to fashion a simple tension member without first devising some end fitting through which the load may be applied; and whether the material be wrought-iron

* For instance, A. G. M. Michell, 'The limits of economy of material in frame structures', *Phil. Mag.* Series 6, *8*, 589 (1904).

or liana, wire rope or string, the stress system in the end fitting is a great deal more complicated than simple tension. There is plenty of scope for theory in the design of tension end fittings, but there is also a great deal of experience; and whether the competition is from the ancient pygmies' mastery of the craft of making knots in lianas, or from Brunel's development of efficient eye bars, experience will often dictate the design. Still the theorist has the final word.

H. L. Cox, *The Design of Structures of Least Weight* (Pergamon, 1965)

If we did not have to consider the effect of end fittings the philosophy of tension structures would be very simple indeed. For one thing, the weight of a tension structure, fitted to carry a given load, would be proportional to its length. That is to say, a rope strong enough to carry a load of one ton over a distance of one hundred metres would weigh just a hundred times as much as a rope safe to carry the same load over one metre. Furthermore, provided that the load were evenly shared, it would make no difference whether a given load were supported by one single rope or tie-bar, or by two ropes or bars each having half the cross-section.

This simple view is upset by the necessity for end fittings: that is to say, by the need to get the load in at one end of the member and out at the other. Even an ordinary rope will need a knot or a splice at each end. The knot or splice will be relatively heavy and may cost money. If we are to do an honest reckoning this weight and cost will have to be added to that of the bare tension member itself. The weight and the cost of the end fittings will be just the same, for a given load, whether the rope be long or short. Thus, other things being equal, the weight and cost of a tension member *per unit length* will be *less* for a long member than for a short one. In other words the weight is *not* directly proportional to the length.

Again, it can be shown, from the algebra and geometry of such a system, that the total weight of the end fittings of *two* tension bars, operating in parallel, is less than that of the end fittings of a *single* rope or bar of equivalent cross-section.* It follows that, in

* Because the cross-section of a tension bar is proportional to the load, whereas the volume of the end fittings increases as the power of 3/2 of the load.

general, weight is saved by subdividing a tensile load between two or more tension members instead of carrying it in a single one.

As Cox points out, the stress distribution in end fittings is always complex and must include more or less severe stress concentrations, from which cracks will spread if they get the chance. Thus both the weight and the cost of the fittings will depend both upon the skill of the designer and also upon the toughness – that is to say, the work of fracture – of the material. The higher the work of fracture, the lighter and the cheaper the fitting will be. However, as we saw in Chapter 5, toughness is likely to diminish as tensile strength increases. In the case of common engineering metals, like steel, the work of fracture falls dramatically with increase of tensile strength.

Thus in choosing a material for a tension member we are commonly faced with incompatible requirements. To reduce the weight of the middle or parallel part of a tie-bar we should like to use a material of high tensile strength. For the end fittings we generally want a tough material – which is only too likely to imply the acceptance of a low tensile strength. Like many difficulties, this one must be solved by a compromise, which in this case depends chiefly on the length of the member. For very long members, such as the wire cables of a modern suspension bridge, it will generally pay to choose a high tensile steel, even if we have to accept extra weight and complication in connection with the end fittings at the anchorages of the cables. After all, there are only two of these, one at each end of the bridge, while there is perhaps a mile of wire in between. Thus the saving of weight over the middle part will more than compensate for any losses at the ends.

But when we come to things like chains with shortish links, the situation is totally different. In each short link the weight of the end fittings may well be greater than that of the middle part and must be carefully considered. This is the case with the supporting chains of the older suspension bridges. Such things were generally made from a tough and ductile wrought iron of quite low tensile strength. As we said in Chapter 10, the tensile stress in the plate links of Telford's Menai bridge chains is less than a tenth of that in the wires of a modern suspension bridge – for this excellent reason. Very similar arguments apply to shell structures such as ships and tanks and boilers and girders which are fabricated from comparatively small plates of iron or steel. It also applies to riveted aluminium structures, such as conventional aircraft. All these may be considered more or less as two-dimensional chains with rather small links. In such cases it *pays* to use a weaker but more ductile material; otherwise the weight of the joints would be prohibitive (see Chapter 5, Figure 13, p. 106).

The multiplication of ropes and wires in ships and biplanes and tents generally results in a saving, rather than an increase, of weight.* Naturally, all this cat's-cradle business incurs the penalty of high wind resistance, high maintenance costs and general complication. This is the price we may have to pay for low structure weight. A similar principle can be seen in animals, where Nature does not hesitate to multiply tension members such as muscles and tendons. Indeed she adopts the same device as the Elizabethan seamen to reduce the weight of end attachments. The ends of many tendons are splayed out into a fan-shaped contrivance which Sir Francis Drake would have called a 'crowsfoot'. Each branch of the tendon has a separate little joint to the bone. Thus the weight (and perhaps the metabolic cost) is minimized.

The relative weights of tension and compression structures

As we saw in the last chapter, the breaking stresses in tension and in compression for a given solid are often different, but for many

*Thinking algebraically, we can put the problem of carrying a load, P, over a length, L, in n parallel tension bars in the form:

$$Z = \rho \frac{P}{s}\left(1 + \frac{k}{WL\sqrt{n}} \cdot \sqrt{\frac{P}{s}}\right)$$

where Z = total weight of all the tension members, per unit length
$\quad\quad\;\; P$ = total load carried
$\quad\quad\;\; s$ = safe working stress
$\quad\quad\;\; k$ = a coefficient connected with the cunning of the designer
$\quad\quad\; W$ = work of fracture of the material
$\quad\quad\;\; n$ = number of tension members employed
$\quad\quad\;\; \rho$ = density of material.

The proof of this is to be found in Cox's *The Design of Structures of Least Weight*. I have modified Cox's formula slightly.

common materials, such as steel, the difference is not very great, and so the weights of *short* tension and compression members are likely to be fairly similar. In fact, because a compression member may not need to have heavy end fittings – whereas a tension member does – a short compression strut may well be lighter, for comparable conditions, than a tension bar.

However, as a strut gets longer, Dr Euler begins to make himself felt. It will be remembered that the buckling load of a long column varies as $1/L^2$ (where L is the length) and this implies that, for a rod of constant cross-section, the compressive strength diminishes very rapidly with increase of length. Thus, to support any given load, a long strut has to be made very much thicker, and therefore heavier, than a short one. As we said in the last section, the same consideration does *not* apply to tension members.

It is revealing to study the problem of carrying one ton (1,000 kg or 10,000 Newtons) over a distance of 10 metres (33 feet) first in tension and then in compression.

IN TENSION. For a steel rod or a cable we might allow a working stress of, say, 330 MN/m² or 50,000 p.s.i. in tension. Taking into account the end fittings, the total weight comes out at about 3·5 kg or about 8 lb.

IN COMPRESSION. To try to carry such a load in compression over such a distance by means of a solid steel rod would be silly, because if a solid rod were thick enough to avoid buckling it would need to be very heavy indeed. In practice we might well use a steel tube, which would have to be about 16 cm (6 inches) in diameter with a wall-thickness of, say, 5 mm (0·2 inch). Such a tube would weigh 200 kg or about 450 lb. In other words it would weigh between fifty and sixty times as much as the tension rod. The cost might well be in the same proportion. Furthermore, if we should want to subdivide a compression structure the situation gets not better but much worse. If we wanted to support a load of one ton, not by a single strut, but by some table-like arrangement of four struts, each 10 metres long, then the total weight of the struts would be twice as great: that is to say, 400 kg or 900 lb. The weight goes on increasing the more the structure is subdivided – in fact as $\sqrt{n}$, where n is the number of columns. (See Appendix 4.)

On the other hand, if we increase the load, keeping the distance the same, then the weight of a compression structure becomes relatively better. For instance, if we increase the load a hundred-fold, that is, from one ton to 100 tons, then, though the weight of a tension member has gone up at least in proportion from 3·5 kg to 350 kg, yet the weight of a single strut to carry this load over 10 metres increases only tenfold, that is, from about 200 kg to

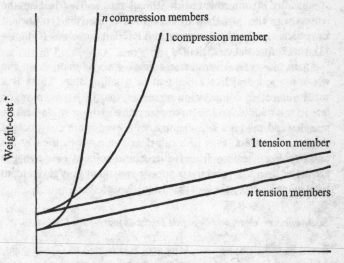

Length, L, over which load has to be carried

Figure 1.   Diagram illustrating the relative weight-cost of carrying a given load over a distance L.

about 2,000 kg. So, in compression, it is proportionately very much more economical to support a heavy load than a light one (Figure 1). All these considerations operate in the same sort of way for panels and shells and plates and membranes as for simple struts and poles and columns (Appendix 4).

Considerations of this kind provide the rationale of things like tents and sailing ships. With such devices it pays, hands down, to collect the compression loads into a small number of masts or poles, contrived to be as short as possible. At the same time the tension loads, as we have said, are better diffused into as many strings and membranes as may be. Thus a bell-tent, which has a single pole but many guy-ropes, is likely to be the lightest 'building' which can be made in proportion to its volume. However, almost any tent will generally be lighter and cheaper than a solid building made from timber or masonry. In the same way, a cutter or a sloop, which has a single mast, is a lighter and more efficient rig than a ketch or a schooner or any other more complicated arrangement with several masts. This is also the reason why the A-shaped or tripod masts used by the ancient Egyptians and by the designers of Victorian ironclads (Chapter 11) were heavy and inefficient.

Again, the typical vertebrate animal, such as man, is on the whole a good deal like a bell-tent or a sailing ship. There is a small number of compression members, that is, bones, more or less in the middle, and these are surrounded by a wilderness of muscles and tendons and membranes – even more complicated than the ropes and sails of a full-rigged ship – which carry the tensions. Furthermore, from the structural point of view two legs are better than four, and the centipede is perhaps only saved from total inadequacy by the fact that its legs are so short.

Scale effects – or second thoughts on the 'square-cube law'

It will be remembered that, long ago, it occurred to Galileo that, whereas the weight of a structure increased as the cube of its dimensions, the cross-sectional area of its load-carrying members increased only as the square, and so the stress in the material of geometrically similar structures ought to increase in direct proportion to the dimensions. Thus a structure which is liable to fail by tensile fracture induced, directly or indirectly, by its own weight must be made of thicker and stockier proportions the larger it becomes. In fact, its members would have to be made disproportionately thicker and heavier than the simple rule would indicate, because there is a sort of 'compound interest' effect. Thus the size of all structures might be expected to be quite strictly limited.

This square-cube law has been bandied about by both biologists and engineers for a long time. Herbert Spencer and, later, D'Arcy Thompson said that it limited the size of animals, such as elephants, and engineers used to explain that it rendered impracticable the building of ships or aircraft appreciably larger than those already in existence. In spite of this, both ships and aircraft continued to get bigger and bigger.

As a matter of fact the square-cube law seems to apply with full force only to the lintels of Greek temples (which are made from weak, heavy stone), icebergs and icefloes (which are made from weak, heavy ice) and things like jellies and blancmanges.

As we have seen, in many sophisticated structures the weight of the compression members is likely to be many times greater than that of the tension parts. Since the compression members are likely to fail by buckling they will become more efficient the larger the load they are called upon to bear – that is to say, the larger the structure is made. For this reason, although there *is* a disproportionate increase of weight with increase of size, the penalty is very much smaller than is implied by the square-cube law. In practice this penalty may be more than offset by various 'economies of scale'. For instance, in a ship or a fish, an aircraft or a bird, the resistance to motion will be nearly in the ratio of the surface area, and this area will diminish, proportionately to the weight, as the size increases. It was Brunel's perception of this which impelled him to design the *Great Eastern*. Brunel's perception was right, though his great ship was a failure, and this is why we build enormous ships, such as super-tankers, today. Furthermore, as we saw in Chapter 5 the size of large animals is more likely to be limited by considerations related to the 'critical Griffith crack length' in their bones than by the square-cube law.

Space-frames versus monocoques

Quite frequently the engineer is faced with a choice between a lattice structure built up, Meccano-fashion, from separate struts and tension rods – which is called a 'space-frame' – and a shell structure in which the load is carried in more or less continuous panels; this is called a 'monocoque'. Sometimes the distinction between the two forms of construction is obscured by the fact that space-frames are covered over with some sort of continuous cladding which does not really carry much load. This is the case with traditional timbered cottages, with modern steel-framed sheds and barns (which are covered with corrugated iron) and, of course, with animals which are covered with shells or scales.

Sometimes the decision about which form to use is dictated by requirements which are not strictly structural. Thus an electricity pylon offers least wind resistance and least area of steel to paint when it is in the form of an open trellis or lattice tower. Again it is generally more convenient to make a water-tank, for instance, from a shell of thickish steel plates than in the form of a trellis supporting a water-tight bag or membrane, even though the latter form may be lighter and is, in fact, the solution usually adopted by Nature for stomachs and bladders.

Sometimes the difference in weight and cost between the two forms of construction is marginal and it may not matter very much which is used. In other cases the difference is very great. As we have seen, a tent is always much lighter and cheaper than any equivalent building made from continuous panels or concrete or masonry. In coachbuilding the old-fashioned 'Weymann' saloon car body, *circa* 1930, which consisted of a wooden space-frame covered with padded fabric, was very much lighter than any of the pressed metal shell bodies which have been used since. In these days of expensive petrol the Weymann body might well be revived.

There is, however, an idea about that monocoque shells are somehow more 'modern' and more advanced than space-frames, which are sometimes considered to be primitive and rather Heath Robinson. Although a good many engineers who ought to know better subscribe to this view, there is in fact no objective structural justification for it. When it comes to carrying loads which are primarily compressive, the space-frame is *always* lighter and usually cheaper than the monocoque. The weight penalty for using a monocoque, however, is less severe when the loads are high in relation to the dimensions, and this, in conjunction with other considerations, may justify the use of shells in some instances. However, for large, lightly loaded structures, such as 'rigid' airships, the space-frame or trellis structure is the only practicable one. The alternative for lighter-than-air transport is not a vast monocoque airship made from an engineer's dream of shiny aluminium plates, but a pressurized bag or 'blimp'.

The transition from the stick and string and fabric construction of the early aircraft to modern monocoques was not dictated by some sudden surge of fashion but was a strictly logical step in aircraft design once certain loads and speeds were reached. As we have said, regarded solely as a means of taking compression and bending, the monocoque is always heavier than the space-frame; but the extra weight required gets less in proportion as the load on the structure increases. On the other hand, regarded as a means of resisting shear and torsion, the monocoque is more efficient than the space-frame.* As aircraft speeds increase, so do the requirements for torsional strength and stiffness. There comes therefore a transition point, which was reached in the 1930s, when it pays, in terms of structure weight, to change over the construction of airframes from space-frame to monocoque. This is especially the case with monoplanes. Thus modern aircraft are usually built as continuous shells, using aluminium sheet, plywood or Fibreglass for the skin. We see an equally logical reversion to space-frame construction in modern hang-gliders, which are very light indeed.

The need to resist large torsional loads is almost confined to artificial structures such as ships and aircraft. As we said in Chapter 12, Nature nearly always manages to avoid torsion, and thus, at least as far as large animals are concerned, monocoques or exo-skeletons are uncommon. Most sizeable animals are vertebrates and therefore highly sophisticated and successful space-frames, not very different in their structural philosophy from biplanes and sailing ships. The avoidance of severe torsional requirements is very noticeable in birds and bats and pterodactyls. It is this which enabled these animals to retain their light space-frame construction when they took to the air. Aircraft designers, please note.

Blown-up structures

It is sometimes interesting to speculate about the technological 'ifs' and 'buts' of history. If Isambard Kingdom Brunel had come

* i.e. for a given cross-sectional area of torsion-box.

upon the railway scene a very few years earlier than he did it is probable that most of the railways of the world would have standardized on a gauge of 7 feet instead of using his rival George Stephenson's 'coal wagon gauge' of 4 feet 8½ inches, which derived from the Roman chariots. The Stephenson gauge has proved something of a handicap, as Brunel predicted it would. If they had a wider gauge today, the railways might perhaps be in a stronger position, technically, and economically, than they are. If so, the world might be slightly different.

On the other hand, if an effective pneumatic tyre had been available around 1830, we might have gone direct to mechanical road transport without passing through the intervening stage of railways at all. In that case the present-day world would have been even more different. In fact the pneumatic tyre was invented about fifteen years too late. It was patented in 1845 by a young man called R. W. Thomson, then aged twenty-three. Thomson's tyre was surprisingly successful technically, but by that time the railways were well established, and the rail interests combined with the horse interests to promote absurd and restrictive legislation, which had the effect of delaying the development of the motor car until the turn of the century.

Since the bicycle was never thought to constitute a serious threat either to trains or to horses its development was legally permitted in Victorian times. The pneumatic tyre was revived with considerable success, for use on cycles, by J. B. Dunlop in 1888. Dunlop made a fortune out of it, but by that time Thomson was dead and his patent had expired. With solid tyres lorries are limited to something like 15 m.p.h., and cars cannot go very much faster. Thomson's invention has not only made fast and cheap road transport practicable; it has also enabled aircraft to operate from dry land. Without pneumatic tyres we should probably have to use some form of seaplane.

Tyres, of course, have the function of spreading and cushioning the load beneath the wheels of a vehicle, and in this they are extremely successful. However, tyres are really only one example of a whole class of blown-up structures. Quite apart from any cushioning effects, blown-up structures provide a very effective way of evading the serious penalties in weight and cost which are incurred when we try to carry light loads over a long distance in bending or in compression. What such a structure does is to carry the compression, not in a solid panel or column which is liable to buckle, but by compressing a fluid, such as air or water. Thus the solid parts have only to sustain tension forces, which, as we have seen, involve very much less weight and cost than compression.

In technology the idea of using blown-up structures in an intelligent way is not new. Around 1,000 B.C., the up-river boatmen of the Tigris and the Euphrates were making boats and rafts from blown-up animal skins. These boats voyaged down-stream carrying, not only produce for sale in the cities of the plains, but also mules or donkeys. On arrival at their destination, the skins were deflated and returned to their home-ports, overland, on the backs of the pack-animals. Nowadays pneumatic boats are common and so are pneumatic tents and furniture. They are often packed up and carried around on cars.

The air-supported roof was invented by the great engineer F. W. Lanchester in 1910. It consists simply of an inflatable membrane, attached at its edges to the ground. It is kept up by air at very low pressure provided by a simple fan arrangement. Although it has to be entered and left by means of an air-lock, this is not usually a very serious handicap in view of the other advantages. Lanchester's roof allows large areas to be covered very easily and cheaply, but its use is at present confined to things like greenhouses and covered tennis courts; it is prevented from being used for factories or houses by rather grandmotherly building regulations.

Of course, one does not have to use air. The sandbag is really another way of doing the same sort of thing, and so are 'Dracone' barges, which are simply large elongated floating sacks, filled with oil or water. They are used on the upper Amazon for transporting oil and are returned, deflated (but not on donkeys), in much the same way as the skin boats of the Euphrates. They are also used to bring fresh water to hotels in the Greek islands for the baths of the tourists.

Blown-up structures probably deserve to be developed much further than they have been for technological uses. However, the great exploiters of this form of construction are plants and animals. Both plants and animals are in business as chemical factories and are, in consequence, full of complicated and messy fluids. Nothing could be more 'natural' and economical, for instance, than to make a worm in the form of an elongated bag stuffed, so to speak, with the worm's squidgy insides.

Clearly, this works very well, and in fact it seems so natural and so economical that one wonders why animals ever bothered to acquire skeletons made from brittle, heavy bones. Would it not be much more convenient, for instance, if men were made like octopuses or squids or elephants' trunks? One view of the question, which was put to me by Professor Simkiss, is that animals never really meant to have skeletons at all; what may have happened was that the earliest bones were simply safe dumping-grounds for unwanted metal atoms in the body. Once animals had produced solid mineral lumps inside their bodies, then they might as well make use of them as attachments for muscles.

Wire wheels

> It won't be a stylish marriage,
> I can't afford a carriage,
> But you'll look sweet upon the seat
> Of a bicycle made for two!

Harry Dacre, *Daisy Bell*

In the traditional wooden carriage wheel the weight of the vehicle is taken in compression by each of the spokes in turn. A carriage is therefore rather like a centipede with a great many long legs which, taken together, are heavy and inefficient. This fact seems first to have dawned upon that remarkable and eccentric man, Sir George Cayley (1773–1857). Cayley was one of the earliest and most brilliant of the aircraft pioneers and he was interested in making better and lighter landing wheels for his aircraft. As early as 1808 it occurred to him that a great deal of weight could be saved by designing wheels in which the spokes were in tension rather than in compression. This thinking led, eventually, to the development of the modern bicycle wheel, in which the wire spokes are in tension while the compressive forces are taken by the rim, which can be quite light and thin, since it is well stabilized against buckling.

In conjunction with the pneumatic tyre, the wire wheel made bicycling practicable for ordinary people – with considerable social consequences, from Daisy Bell onwards. The saving in weight is, however, mostly confined to large lightly loaded wheels, such as bicycle wheels. When the wheel becomes smaller and the loads larger there is generally not much advantage to be gained from using tension spokes. In modern sports cars pressed-steel wheels are very little heavier than wire wheels, which are usually not worth the bother and expense.

On choosing a better material – and what is a 'better' material anyway?

Nature may be supposed to know her business when she chooses between the various possibilities in the way of biological tissues; but mere men, even very great men, seem to have the strangest ideas about materials. According to Homer, the bow of Apollo was made of silver* – a metal whose strain energy storage is negligible. In a rather later age we were told that the floors of Heaven were made of gold, or alternatively of glass: both very unsuitable substances. Poets are always quite hopeless about materials; but most of the rest of us are not much better. In fact very few people ever think rationally about the subject at all.

Quirks of fashion and prestige seem to play a large part in the matter. Gold is not really a very good material for watches, nor is steel for office furniture. The Victorians insisted on making all sorts of improbable articles, such as umbrella stands, out of cast iron, and there is the story of the African chief who had his palace made out of the same substance.

Although the choice of materials is sometimes irrational and eccentric, more often it is highly traditional and conservative. Of course there is sound reason behind a good deal of traditional materials selection, but it is so mixed up with unreason that it is

---

* *Neque semper arcum tendit Apollo*! ('Neither is Apollo perpetually drawing his bow', Horace, *Odes* II, x, 19). Horace perhaps knew that silver creeps nearly as badly as lead.

difficult to separate the two. Artists from Lewis Carroll to Dali have discovered that it is possible to impart a considerable psychological shock merely by implying that some familiar object might be made from an apparently unsuitable material, such as rubber or bread and butter. Engineers are very susceptible to these effects; they would be a good deal shocked nowadays at the idea of a large wooden ship. Our ancestors were much more shocked at the idea of an iron one.

The acceptability of various materials changes with time in curious and interesting ways. Thatch is a case in point. Thatch was once the cheapest and least regarded of roofing materials, but in the poorer country districts it often had to suffice even for the roofs of churches. During the eighteenth century, when these parishes became richer, subscriptions were raised to replace the thatch by slates or tiles. Sometimes the money was inadequate to do the whole job, and in these cases the thatch had perforce to be left on those areas of the church roof which were not likely to be seen by passers-by; only the side which faced the main road was tiled. Nowadays the balance of prestige is reversed, and in the Home Counties thatched roofs are the pride and joy of the wealthier of the business fraternity.

Materials, fuel and energy

The twentieth century may be known to posterity as the 'age of steel and concrete'. It may also be known as the 'age of ugliness', and perhaps by other unpleasant names as well, such as the 'age of waste'. It is not only engineers who are obsessed by steel and concrete (and quite indifferent to appearances); politicians and the man in the street seem to have caught the same infection. The disease seems to have originated two hundred years ago with the Industrial Revolution and cheap coal – which led to cheap iron – which led to iron steam engines fitted to turn that coal into cheap mechanical energy: and so on round and round in ever more energy-intensive circles. Thus coal and oil store a great deal of energy packed into a small volume. Engines process a great deal of this energy very quickly and within a small space. They then deliver the energy as electricity or mechanical work in concentrated forms. On this concentration of energy our whole contemporary technology rests. The materials of this technology, steel, aluminium and concrete, themselves require a great deal of energy to manufacture them; how much energy is indicated in Table 6. Because they need so much energy to make them these

TABLE 6

Approximate energies required to produce various materials

Material	η = energy to manufacture Joules $\times 10^9$ per ton	Oil equivalent tons
Steel (mild)	60	1·5
Titanium	800	20
Aluminium	250	6
Glass	24	0·6
Brick	6	0·15
Concrete	4·0	0·1
Carbon-fibre composite	4,000	100
Wood (spruce)	1·0	0·025
Polyethylene	45	1·1

Note. All these values are very rough and no doubt controversial; but I think that they are in the right region. The value given for carbon-fibre composites is admittedly a guess; but it is a guess founded upon many years of experience in developing similar fibres.

materials can only be employed with profit within an energy-intensive economy. We are not only investing money capital in a technical device; we are also investing energy capital, and in both cases it is necessary to secure a fair return on the investment.

In spite of the high cost and increasing scarcity of energy the trend in the energy-intensive direction is increasing rather than diminishing. Advanced engines, such as gas turbines, process more and more energy, more and more hectically, within less and less space. Advanced devices require advanced materials, and the newer materials, such as high-temperature alloys and carbon-fibre plastics, consume more and more energy in their manufacture.

Most probably this kind of thing cannot go on for very much longer, for the whole system is entirely dependent upon cheap and concentrated sources of energy, such as oil. Living Nature may be regarded as an enormous system for extracting energy, not from concentrated but from diffuse sources, and then using that energy with the uttermost economy. Many attempts are on foot at present to collect energy for technology from diffuse sources, such as the sun, the wind or the sea. Many of these are likely to fail because the energy investment which will be needed, using conventional collecting structures built of steel or concrete, cannot yield an economic return. A quite different approach to the whole concept of 'efficiency' will be needed. Nature seems to look at these problems in terms of her 'metabolic investment', and we may have to do something of the same kind.

It is not only that metals and concrete require a great deal of energy, *per ton*, to manufacture (Table 6), but also that, for the diffuse or lightly loaded structures which are usually needed for systems of low energy intensity, the actual weight of devices made from steel and concrete is likely to be very many times higher than it would be if we used more sensible and more civilized materials.

As we shall shortly see, timber can be one of the most 'efficient' of all materials in a strictly structural sense. For large dimensions and light loads, a wooden structure is many times lighter than one made from steel or concrete. One of the difficulties with timber, in the past, has always been that trees take a long time to grow and wood is slow and expensive to season.

Probably the most important development in materials during the last few years has been that made by the plant geneticists who have been breeding fast-growing varieties of commercial timbers. Thus varieties of *Pinus radiata* (Weymouth pine) are now being planted which, in favourable conditions, will increase in diameter by up to 12 centimetres per year and may be fit for felling, as mature timber, in six years. So there is a good prospect of timber becoming a crop which can be grown on a short time-cycle. Nearly all the energy which is needed to make it grow is provided, free, by the sun. Presumably, when one has finished with a timber structure, it could be burnt to yield up most of the energy which it has collected while it was growing. This is, of course, in no way true of steel or concrete.

Again, timber used to need lengthy and expensive seasoning in heated kilns, which used up a good deal of energy. As a result of recent research it is now possible to season sizeable soft-wood scantlings in twenty-four hours, at a very low cost. These are very important developments in relation to structures and to the world energy situation, and it behoves us to take account of them.

Some algebraical analysis of the structural efficiencies in various roles and in terms of weight, of different materials, is given in Appendix 4. The design of a number of high-technology structures, such as aircraft, is largely controlled by the criterion E/ρ: that is to say, by the 'specific Young's modulus' which governs the weight-cost of the overall deflections. It happens that, for the majority of traditional structural materials, molybdenum, steel, titanium, aluminium, magnesium and wood, the value of E/ρ is sensibly constant. It is for this reason that, over the last fifteen or twenty years, governments have spent such large sums of money in developing new materials based on exotic fibres such as boron, carbon and silicon carbide.

Fibres of this sort may or may not be effective in aerospace; but what seems to be certain is that not only are they expensive but they also need large amounts of energy to make them. For this reason their future use is likely to be rather limited, and, in my own view, they are not likely to become the 'people's materials' of the foreseeable future.

TABLE 7

The efficiency of various materials in different roles

Material	Young's modulus, $E\,\mathrm{MN/m^2}$	Specific gravity, ρ grams/c.c.	E/ρ	$\dfrac{\sqrt{E}}{\rho}$	$\dfrac{\sqrt[3]{E}}{\rho}$
Steel	210,000	7·8	27,000	59	7·7
Titanium	120,000	4·5	27,000	77	11·0
Aluminium	73,000	2·8	26,000	99	15·0
Magnesium	42,000	1·7	25,000	120	20·5
Glass	73,000	2·4	30,000	114	17·5
Brick	21,000	3·0	7,000	48	9·0
Concrete	15,000	2·5	6,000	49	10·0
Carbon-fibre composite	200,000	2·0	100,000	225	29·0
Wood (spruce)	14,000	0.5	28,000	240	48·0

The requirement for a strict and expensive control of overall deflections is likely to be a very limited one; however, as we have seen, the weight-cost – and often the money cost – of carrying compressive loads is frequently very high. The weight-cost of carrying a compressive load in a column is governed, not by E/ρ, but by $\dfrac{\sqrt{E}}{\rho}$. The weight-cost of a panel is controlled by $\dfrac{\sqrt[3]{E}}{\rho}$ (Appendix 4). These requirements are summarized in Table 7. It will be seen that there is a large premium on low density; thus steel comes out rather badly, even compared with bricks and concrete. Furthermore, for many light-weight applications – such as airships or artificial limbs – wood is even better than carbon-fibre materials, besides being much cheaper.

In Table 8 these virtues are expressed in terms of energy-cost.

TABLE 8

The structural efficiency of various materials in terms of the energy needed to make them

Material	Energy needed to ensure a given stiffness in the structure as a whole	Energy needed to produce a panel of given compressive strength
Steel	1	1
Titanium	13	9
Aluminium	4	2
Brick	0·4	0·1
Concrete	0·3	0·05
Wood	0·02	0·002
Carbon-fibre composite	17	17·0

These figures are based on mild steel as unity. They are only very approximate.

Here the advantage of the traditional materials – wood, brick and concrete – is overwhelming. This table makes one wonder whether the pursuit of materials based on exotic fibres is really justified. What really pays off for most of the common purposes of life is not carbon fibres, but holes. Nature tumbled to this a long time ago when she invented wood; and so did the Romans when they started to build churches from empty wine bottles. Holes are enormously cheaper, both in money and in energy, than any conceivable form of high-stiffness material. It would probably be better to spend more time and money on developing cellular or porous materials and less on boron or carbon fibres.

Chapter 15   A chapter of accidents

– a study in sin, error and metal fatigue

> *Have you heard of the wonderful one-hoss shay*
> *That was built in such a logical way,*
> *It ran a hundred years to a day,*
> *And then, of a sudden, it –*

Oliver Wendell Holmes, *The One-Hoss Shay*

The entire physical world is most properly regarded as a great energy system: an enormous market-place in which one form of energy is for ever being traded for another form according to set rules and values. That which is energetically advantageous is that which will sooner or later happen. In one sense a structure is a device which exists in order to delay some event which is energetically favoured. It is energetically advantageous, for instance, for a weight to fall to the ground, for strain energy to be released – and so on. Sooner or later the weight *will* fall to the ground and the strain energy *will* be released; but it is the business of a structure to delay such events for a season, for a lifetime or for thousands of years. All structures will be broken or destroyed in the end – just as all people will die in the end. It is the purpose of medicine and engineering to postpone these occurrences for a decent interval.

The question is: what is to be regarded as a 'decent interval'? Every structure must be built so as to be 'safe' for what may reasonably be considered an appropriate working life. For a rocket case this might be a few minutes, for a car or an aircraft, ten or twenty years, for a cathedral perhaps a thousand years. Oliver Wendell Holmes's 'one-hoss shay' was constructed to last for a hundred years – neither more nor less – and it disintegrated, exactly as planned, on 1 November 1855, just as the parson had reached 'fifthly' in the composition of his sermon. But, of course, this was nonsense. Again, the egregious but heroic Mr Honey in Nevil Shute's *No Highway* predicts the failure of the tail of the Reindeer airliner from 'metal fatigue' after exactly 1,440 flying hours – plus or minus a day or so. This again was nonsense, as Nevil Shute must have known, being an experienced aircraft designer.

It is impossible, in practice, to plan for a 'safe' life of exactly so many hours or years. We can only consider the problem in statistical terms and in the light of accumulated data and experience. We then build in whatever margin of safety seems reasonable. All the time we are working on a basis of probabilities and estimates. If we make the structure too weak we may save weight and money, but then the chance of the thing breaking too soon will become unacceptably high. Contrariwise, if we make a structure so strong that, in human terms, it is likely to last 'for ever' – which is what the public would like – then it will probably be too heavy and expensive. As we shall see, there are many cases where more danger is incurred by extra weight than is avoided by the corresponding increase of strength. Because we are necessarily working on a statistical basis, when we design a practical structure for a realistic life we have to accept that there is always some finite risk, however small, of premature failure.

As Sir Alfred Pugsley points out in his book *The Safety of Structures*,* it is just at this rather interesting stage that we may have to abandon a strictly logical approach to the problem. As Pugsley says, the human emotions are quite exceptionally sensitive to the fear of structural failure, and the layman clings with great tenacity to the idea that any structure or device with which he is personally associated should be 'unbreakable'. This crops up in all sorts of connections; sometimes it does no harm, sometimes the effect is counter-productive. During the last war aircraft designers had the choice, to some extent, of trading off structural safety against other qualities in the aircraft. Now the losses of bomber aircraft by enemy action were very high, something like one out of twenty in each sortie.† Against this, the losses from structural failures were very few, much less than one aircraft in

* Arnold, 1966.
† Each 'tour of duty' for an airman in Bomber Command consisted of thirty sorties or operational flights. Such service was therefore exceptionally dangerous. The loss of life in Bomber Command was comparable to that of the German U-boat crews, which was notoriously high.

ten thousand. The structure of an aeroplane accounts for practically a third of its total weight, and it would have been rational to have slimmed the structural parts of the bombers in return for other advantages.

If this had been done there would have been some small increase in the structural accident rate, but the weight that would have been saved could have been invested in more defensive guns or in thicker protective armour. In that case there would no doubt have been a significant reduction in the net, or overall, casualty rate. But the airmen would not hear of anything of the kind. They preferred the big risk of being shot down by the enemy to the smaller risk of the aircraft breaking up in the air for structural reasons.

Pugsley suggests that the feeling that it is in some way outrageous for a structure to break may be inherited from our arboreal ancestors, who were frightened, above all things, that the trees in which they lived might break beneath them – when down would come baby and cradle and all. And besides, the ancestors and their babies would fall into the mouths of their enemies on the ground, such as sabre-toothed tigers or whatnot. Whether this is the real reason or no, engineers have to take these sort of feelings into account, even though the extra weight incurred may involve dangers of its own.

The accuracy of strength calculations

It is implicit in any rational approach to questions of strength and safety that the engineer should be able to predict, with sufficient accuracy, the strength of a proposed structure when it is new – even if he is in doubt about how long it may be expected to last. While this may be roughly the case for simple structures such as ropes and chains and straightforward beams and columns, as we saw in Chapter 4, it is just not true at all for the more elaborate and critical artefacts, things like aircraft and ships.

Since there is available a great body of accumulated experience with various kinds of structures, since there also exists a vast and highly mathematical literature on the subject, and since academic elasticians, in their pride, deliver endless lectures about the theory of structures, that statement might be regarded as sticking one's neck out. However, it is true.

Consider, for instance, the statistics for the strength of aircraft. Since the saving of weight is important and since the consequences of failure are very horrible, a great deal of care and thought is naturally given to the structural design of aeroplanes, and every detail is meticulously checked. The drawings and calculations are made by highly skilled designers and stressmen and draughtsmen, using the most scientific methods. When these people have done their sums the strength calculations are checked, quite independently, by an entirely different set of experts. Thus the strength predictions which are finally arrived at are about as accurate and painstaking as is humanly possible. Finally, and to make quite certain, an actual full-scale airframe is tested to destruction.

It is not possible to give really up-to-date results because so few different types of aeroplane have been ordered in recent years that the figures are not statistically significant. However, when aircraft were simpler and cheaper, a comparatively large number of designs reached at least the prototype stage. Between 1935 and 1955 something in the region of a hundred different kinds of aeroplane were built and tested to destruction in this country. Thus the results for this period form a fairly reliable guide with some sort of statistical basis.

Naturally, the actual figures of the required strengths for these different aircraft varied a great deal, according to the size and type of aeroplane. However, each design team could be said to be aiming at that strength which is known in the jargon of the aircraft trade as '120 per cent fully factored load'.* If structural design were anything like an exact profession one would expect the various test results, when plotted on a curve or 'histogram', to cluster pretty closely around the value for 120 per cent fully factored load, give or take a very little. In other words the results should produce a narrow 'normal' or bell-shaped distribution curve, much like Figure 1.

As is fairly well known, nothing of the sort happened. When the results are plotted the histogram looks more like Figure 2. The

* The extra 20 per cent was required by the airworthiness authorities so as to cater for variations in the material and in the assembly procedures.

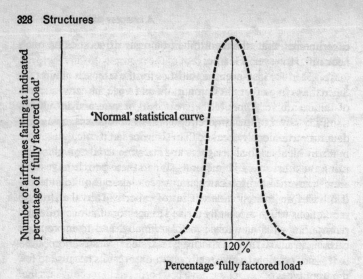

Figure 1. Expected statistical distribution of experimental aircraft strengths (schematic diagram).

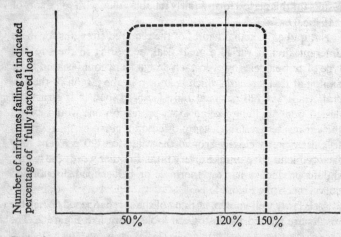

Figure 2. Actual distribution of strengths of airframes broken in test-frame, 1935–55 (very approximate schematic diagram).

experimental strengths tend to be randomly distributed between about 50 per cent and 150 per cent of the required or fully factored load. That is to say, even the most eminent designers cannot be relied upon to predict the strength of an aeroplane within a range of three to one. Some of these aircraft were less than half as strong as they should have been; others were much too strong and therefore considerably heavier than they needed to be.

When it comes to ships, there are really no data on which one can base this sort of judgement – for the reason that ships are almost never tested to destruction under laboratory conditions. It is therefore impossible to tell how good or bad naval architects are at their job – at least as far as strength predictions are concerned. However, as we said in Chapter 5, the number of structural accidents to ships is considerable, and it seems very possible that the number of accidents per ton-mile is increasing at the present time.

With regard to bridges, the problem of strength calculation is in some respects easier than with ships and aircraft, since the loading conditions are less varied. Nevertheless, the number of failures in modern bridges is quite significant.

Designing by experiment

> Now, in building of chaises, I tell you what,
> There is always somewhere a weakest spot –
> In hub, tire, felloe, or spring or thill,
> In panel, or crossbar, or floor, or sill,
> In screw, bolt, thoroughbrace – lurking still,
> Find it somewhere you must and will.

Oliver Wendell Holmes, *The One-Hoss Shay*

The fallibility of the theoretical design process is, of course, the reason for the insistence on the experimental strength testing of all aircraft. However, the benefits of an experimental approach extend still further. We have assumed that it ought to be the designer's aim for a structure to fail, the first time it is tested, exactly at the required load. But even the most scientifically designed structure is very unlikely to be of consistent strength throughout all its parts – like the legendary shay, where

> *... the wheels were just as strong as the thills,*
> *And the floor was just as strong as the sills,*
> *And the panels were just as strong as the floor –*

and so on for many components and many lines of verse.

On the test-frame the structure breaks at the weakest place; all the rest of the structure is therefore of greater strength. If an airframe fails initially at just the required 120 per cent it follows that much the greater part of the structure is too strong for its purpose, and this extra strength is completely wasted. But we have no means of knowing where and how to lighten the structure. Repeated tests on large structures are expensive and time-consuming, but, where time and money allow, it is better to arrange, if possible, for the initial failure to occur at a load comfortably below the official 120 per cent. The weak place thus indicated can then be strengthened and the whole structure retested – and so on.

The war-time Mosquito bomber, which was one of the most successful aircraft in history, failed initially at 88 per cent of the factored load – in the rear wing-spar. The aeroplane was then progressively strengthened up to a figure of 118 per cent. It was owing, partly, to the exceptionally light and strong airframe that the performance of this aircraft was outstanding.

This is, roughly speaking, the Darwinian method, which Nature seems to rely on to develop her own structures – though she seems to be in less of a hurry and less mindful of the value of life than are most civilized human engineers. It is also, to a notable extent, the method employed by the makers of cars and other cheap, mass-produced goods. These people tend to make their products deliberately too weak for their purpose and to rely upon customers' complaints to detect the significant faults.

Thus a great deal of the strength-predicting element of design boils down to a sort of game in which we try to spot the weakest link in a load-bearing system. The more complicated the structure, the more difficult and unreliable this becomes. Fortunately, the design of a great many structures, from furniture and buildings to aeroplanes, is rescued from becoming a completely ridiculous process by the fact that the stiffness requirements may be more exacting than the strength requirements. Thus, if the structure is made stiff enough for its purpose, it may then very well be sufficiently strong. Since the deflections in a structure depend upon its general character rather than upon the existence of a 'weakest link', stiffness predictions are much easier to make, and more reliable, than strength predictions. This is what we really mean when we talk about designing a thing 'by eye'.

How long will it last?

> *This also said Phocylides:*
> *A tiny rock-built citadel*
> *Is finer far, if ordered well,*
> *Than all your frantic Ninevehs.*

Phocylides (translated by Sir Maurice Bowra)

In discussing the strength and stability of the masonry cathedrals Professor Jacques Heyman has laid down the principle that 'If a structure will stand for five minutes, it will stand for five hundred years.' For masonry structures built upon rock this is, broadly speaking, true. However, many cathedrals and other buildings have been founded upon soft ground. If this soft soil creeps (Chapter 7) – which happens quite often – curious things will happen, such as the Leaning Tower of Pisa. Such movements take time and can often be predicted, but they are very expensive to put right, and a certain number of buildings, both ancient and modern, fall down or have to be demolished for this reason.

In most types of structure, rot and rust are very active agents of decay. It is partly the fear of rot which has turned engineers and architects in Britain against timber. However, the poor benighted foreigners in America and Canada and Scandinavia and Switzerland, who build between them about 1,500,000 wooden houses each year, do not seem to be troubled with rot to the same extent, and it might be a good idea to see how they manage these things. The use of wood is greatly on the increase in these countries.

Timbers vary a great deal in their natural resistance to decay, and Lloyd's Register allocates a fixed number of years of life to each of the different timbers which are used in shipbuilding. However, with modern knowledge and methods of treatment, it should be possible to get a practically indefinite life from almost any kind of wood.

Most metals corrode in service. Modern mild steel rusts very much worse than Victorian wrought iron or cast iron, and so rust is, to some extent, a modern problem. Because the cost of labour is high, the cost of the painting and maintenance of steelwork is high. This is one good reason for using reinforced concrete, since steel embedded in concrete does not rust. In fact large modern ships, such as tankers, are constructed for a life of about fifteen years; on the whole it is cheaper to scrap than to paint. The life of cars is even shorter, usually for the same reason. It is true that for some structures one could use stainless steel but it is by no means always proof against corrosion, and stainless steels are expensive and awkward to fabricate. Besides this, the 'fatigue properties' of stainless steels are usually bad.

These are some of the reasons for choosing aluminium alloys; but, apart from the extra cost, there are a number of cases where the stiffness of aluminium has proved inadequate. The difficulty of welding aluminium is also a handicap. Some Communist countries see a great future for aluminium and have invested largely in aluminium plants. The London stock-markets were considerably shaken by the Tube Investments–British Aluminium take-over bid in 1961. However, the market for aluminium has not expanded to anything like the extent which was anticipated by the business-men concerned in this transaction. In any case it requires more energy to make aluminium than to make steel.

Even if the material of a structure does not deteriorate, its life may be subject to statistical effects which are sometimes cal-culable – and sometimes not. Many structures are likely to be broken only in rather exceptional circumstances, and it may be a long time before these circumstances arise. Freakishly high waves, in the case of a ship, and exceptionally severe upward gusts with aircraft are cases in point. Some structures are likely to be broken only by unusual combinations of events. For a bridge this might be the coincidence of very high winds with exceptional traffic loads. Although such eventualities ought to be provided for, it may be many years before they actually happen. So an essentially unsafe structure may stand for a long time, simply because it has never been fully tried.

Responsible engineers do, of course, try to predict things of this sort and to make structural provision for them, but in many cases such peak loads shade off into what the insurance companies call 'acts of God'.* If a ship runs into a large bridge, destroying both the bridge and the ship, as happened recently in Tasmania, it is very difficult to see what either the naval architect or the bridge designer could have been expected to do about it from the structural point of view. The problem is one not for the structural engineer but for the local Pilotage Association. Again, aircraft cannot be designed to be flown into mountains. We do, to a certain extent, design cars to be driven into brick walls without killing the passengers, but then we do not expect the car to be of much use afterwards.

Metal fatigue, Mr Honey and all that

One of the most insidious causes of loss of strength in a structure is 'fatigue': that is to say, the cumulative effect of fluctuating loads. The dramatic possibilities of fatigue in metals were first exploited in popular literature in 1895 in Kipling's account of what happened when the propellor of the *Grotkau* dropped off somewhere in the Bay of Biscay because of a fatigue crack in the tailshaft.† Kipling went out of fashion, but public interest in fatigue was revived in 1948 by Nevil Shute's *No Highway*. The success of this story, both as a book and as a film, was no doubt partly due to the character of Mr Honey, the archetypal boffin, but perhaps still more to the three Comet disasters, which occurred not very long afterwards. As Whistler remarked some time ago, Nature keeps creeping up on Art. The circumstances of the Comet accidents were not very different from those imagined in *No Highway*, except that many more lives were lost and a great deal of damage was done to the British aircraft industry.

As a matter of fact, engineers' knowledge of fatigue effects in metals goes back rather over a hundred years. Indeed it was not long after the Industrial Revolution that it began to be noticed that the moving parts of machinery would sometimes break at loads and stresses which would have been perfectly safe in a

* An act of God has been defined by A. P. Herbert as 'That which no reasonable man would expect'.
† *Bread upon the Waters* (published in *The Day's Work*).

stationary component. This was especially dangerous in railway trains, whose axles would sometimes break off suddenly and for no apparent reason after they had been in service for a time. The effect soon came to be known as 'fatigue', and the classical researches on the subject were carried out during the middle years of the nineteenth century by a German railway official called Wöhler (1819–1914). From his photograph Herr Wöhler looks exactly what one would expect a German nineteenth-century railway official to look like; but he did a very useful job.

As we said in Chapter 5, even though there may be a high local stress at the tip of a notch or a crack, the crack will not extend – so long as it is shorter than the 'critical Griffith length' – because making it spread requires work to be done against the 'work of fracture' of the material. However, when the stress in the material is a fluctuating one, slow changes take place within the crystalline structure of the metal, and this is particularly likely to happen in the region of a stress concentration. These changes have the effect of reducing the work of fracture of the metal in such a manner that the crack is able to extend, very slowly, even though it may be much shorter than the 'critical length'.

In this way a tiny unseen crack may start from any hole or notch or irregularity in a stressed metal and may spread across the material, which is not, as a whole, changed in any obvious way. Sooner or later, such a 'fatigue crack' will reach the critical length for an ordinary common or garden crack. When this happens, the crack will immediately speed up and run right across the material, often with very serious consequences. It is usually quite easy to diagnose a fatigue crack after failure because of its characteristic striped or banded appearance. Before rupture, however, an incipient fatigue failure may be practically impossible to spot.

Naturally metallurgists and others do a great deal of experimental fatigue testing on their materials, and a great many different types of testing machine are now available for the purpose. It is common to consider the fatigue properties of a metal in terms of a reversed stress ($\pm s$) – that is to say, the sort of stress which would occur in a rotating cantilever, such as the axle of a vehicle. (There are ways of converting these results to other conditions of fluctuating stress.) This reversed stress ($\pm s$) is usually plotted on a graph against the logarithm of the number (n) of times the stress has to be applied to a specimen to cause failure. This is sometimes called an 's–n diagram'.

The s–n diagram for a typical steel would look like Figure 3. It will be seen that the 'curve' is a dog-legged affair which flattens

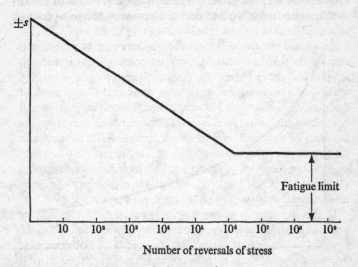

Figure 3. Typical fatigue curve for iron or steel.

off after about a million reversals – which might be equivalent to about 3,000 miles of service for the axle of a car or train, or about ten hours of running for an ordinary car engine, which, of course, goes round much faster than the wheels. The existence of a definite 'fatigue limit' of this nature for materials like iron and steel constitutes a great comfort to the engineer. If his engine or his vehicle will run for 10^6 or 10^7 revolutions – which may only take a few hours – then there is some hope of its being safe indefinitely. But fatigue is a danger which always needs to be considered.

Aluminium alloys do not have a definite fatigue limit but tend to tail off, something after the fashion of Figure 4. This makes them more dangerous to use and accounts for some apparently old-fashioned prejudices in favour of steel for use in machinery and other structures.

The Comet accidents, which occurred in 1953 and 1954, naturally caused consternation and well-justified alarm. The investigation of these accidents by Sir Arnold Hall and a large team of experts was a classical feat, not only of engineering detection, but also of deep-sea salvage. The broken parts of one of the aircraft, which had fallen into the Mediterranean, had to be dredged

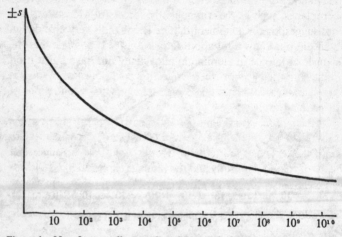

Figure 4. Non-ferrous alloys such as brass and aluminium frequently do not show any definite fatigue limit.

up from a depth of over 300 feet or 50 fathoms. The salvage people managed to recover practically the whole of the aeroplane and the innumerable fragments covered the floor of a large hangar at Farnborough. As far as I remember, no piece was more than two or three feet across.

The Comet was one of the earliest airliners to have a pressurized fuselage. The main purpose of this was, of course, to spare the passengers from the discomfort and danger of the atmospheric pressure changes associated with change of altitude. In the old days, when flying over the Rocky Mountains, one used to have to eat one's lunch while wearing an oxygen mask: this now ranks as one of those lost skills. In a pressurized aircraft the fuselage becomes, in effect, a cylindrical pressure vessel, not unlike a very thin-walled boiler, which is pressurized and relaxed every time the aircraft climbs and descends.

The lethal mistake in the design of the Comet lay in not realizing sufficiently the danger of 'fatigue' occurring at stress concentrations in the metal of the fuselage under these circumstances. The Comet was built from aluminium alloys, and most of de Havilland's previous experience had been gained with wooden aeroplanes, such as the triumphantly successful Mosquito. I am not suggesting for a moment that de Havilland's very able design staff did not know a lot about fatigue; but it is possible that the danger of fatigue in aluminium alloys may not have burnt itself sufficiently deeply into their collective consciousness. Wood is much less susceptible to this danger than metals – which is one of its great advantages.

In each of these accidents cracks seem to have started from the same small hole in the fuselage and spread, slowly and undetected, until they reached the 'critical Griffith length'. Whereupon the skin tore catastrophically and the fuselage exploded like a blown-up balloon. By repeatedly pressurizing a Comet fuselage in a large tank of water at Farnborough, Sir Arnold Hall was able to reproduce the effect so that it could be observed, as it were, in slow motion.

Part of the trouble about the Comet accidents was that the fatigue cracks which must have existed were never spotted by an inspector, perhaps because he was not expecting to find them, but more probably because they were too short to be seen easily. Nowadays aircraft fuselages are designed to contain with safety cracks up to about two feet long, and one would think that so long a crack could hardly fail to be seen in good time. There is, however, the story about the two cleaners at London Airport. These ladies finished sweeping out the cabins of an empty airliner late one night. They shut the door and went down the steps on to the tarmac.

'You've forgotten to switch off the light in the toilet, Mary.'

'How do you know?'

'Can't you see it shining through the crack in the fuselage?'

Accidents to wooden ships

Before the days of railways nearly all the heavy traffic went by water. Besides the deep sea trade and the Continental trade and the inland trade by river and canal, there was an even larger coastal trade. Many thousands of little wooden brigs and schooners, of the kind caricatured by W. W. Jacobs, transported anything and everything, not only into the coastwise creeks and harbours, but to almost every possible or impossible beach. A ship would be grounded on the beach at high water and, when the tide fell, would unload her coal or bricks or lime or household furniture into carts which were driven alongside. When the tide rose again, she would slip away to sea and then go and do it all over again somewhere else.

Naturally this was rather a risky business, but during the eighteenth century most of the smaller vessels could afford to lay up and refit during the worst of the winter – when the crews could see something of their families and of the local pubs. This slightly idyllic and not exceptionally dangerous state of affairs was upset by the more competitive conditions of the nineteenth century. Under commercial pressures vessels had to trade throughout the winter and could not afford, as a rule, to wait for weather. Indeed the regularity of some of these little sailing ships would put a lot of modern goods trains to shame.

But, of course, a price had to be paid. During the middle 1830s an average of 567 shipwrecks occurred round the coasts of this country each year; as a result, a yearly average of 894 lives were lost. Whether these figures are better or worse, per ton-mile of goods delivered, than modern lorries I do not know. At any rate the public conscience was disturbed at the time and Parliament appointed a Select Committee to investigate the 'Causes of Shipwrecks'. After hearing a great deal of evidence, the committee reported that, apart from minor causes, shipwrecks in this country could principally be attributed to the following conditions in ships:

1. Defective construction.
2. Inadequacy of equipment.
3. Imperfect state of repair.

They pronounced 'That the defective construction of ships appears to have been greatly encouraged by the system of classification [i.e. the rules governing construction and repair of insured ships] which, from the year 1798 up to 1834, was followed by Lloyds.'

The committee went on to add that the system by which the government measured ships for tonnage dues encouraged thoroughly unseaworthy shapes of hull. The bureaucratic mind does not seem to change very greatly through the centuries.

To be fair, the problem of framing regulations for the strength and safety of ships, or any other kind of structure, is an extraordinarily difficult one. No doubt a certain amount of progress has been made in the matter since the 1830s. At the same time, and in a different sense, a great deal of technical progress has been prevented – especially by the various building regulations. As Pugsley points out in *The Safety of Structures*, it is inherently impossible to make regulations about the strength of structures which are proof against both fools and knaves without preventing, or at best handicapping, development and innovation. Regulations for structural safety are presumably necessary, but some of them are not only stultifying; they can be the actual cause of accidents.

To return to wooden ships: not only the clippers but the little brigs and brigantines and topsail schooners and barges – which were so beautiful and so satisfying – have all gone, and the yards that used to build them are now turning out yachts. The structural problem of a wooden yacht is both more and also less severe than that of larger vessels. Yachts' hulls are not bumped on shingle beaches while carrying cargoes of stone or coal, but they have a more difficult problem with regard to local impacts which their thin skins are not well fitted to resist.

Now that long voyages in small yachts have become so fashionable this question of the impact strength of the hull has become important. Yachts voyaging in deep waters have repeatedly been attacked and sunk by killer whales. These animals weigh about six tons and swim at around thirty knots. They seem to have a special hatred for small yachts, which they ram and hole below the waterline. This has now happened so often that the possibility cannot any longer be classed as an 'act of God' (Poseidon presumably) but is a serious hazard which must be guarded against.

It is probably impracticable to make the sides of a small yacht thick enough and strong enough to resist such an attack. The best thing to do would seem to be to provide some sort of inflatable floatation gear to keep the yacht afloat – and preferably sailable – after she has been holed. So far, those who have survived these attacks have done so by taking to the dinghy, which, naturally, gave most of them a very unpleasant time before they were picked up by a steamer after many days or weeks.

More about boilers and pressure vessels – and something with boiling oil in it

For a considerable number of years before the railway system was completed much of the passenger and express freight traffic was carried by steamship. During the first half of the nineteenth century, not only were there far more steamers running to more Continental ports than is the case today, but there were also very numerous services between towns in Great Britain. Considerably the cheapest – and often the quickest and most comfortable – route from London to such places as Newcastle, Edinburgh or Aberdeen was by steamboat.

Accidents were fewer in steamships than in sailing vessels only because there were many fewer steamships. Nevertheless, between 1817 and 1839, there were ninety-two major accidents to steamships in British waters. Of these, twenty-three were due to boiler explosions. This is nothing like as bad a record as that of the American river steamers a few years later; but it is quite bad enough.

Some of the early boilers were made from unsuitable materials, such as cast iron. At least one cast-iron boiler, that of the S.S. *Norwich*, duly burst and killed several people. Even when boilers were more or less properly constructed from wrought iron, they were very commonly neglected and allowed to rust through until they burst. This was the cause of the loss of the *Forfarshire* on the

Farne Islands in 1838. Five people were rescued by Grace Darling's superb feat of seamanship.*

Again a Parliamentary Committee was appointed, which reported in 1839 and produced an extensive, thorough, factual and almost incredible document. During the boom years of the expansion of the steam engine, sober, let alone competent, responsible or intelligent engine-room staff were almost unobtainable, even at very high wages. These people treated their engines and boilers with a degree of ignorance and carelessness which almost passes belief. For instance:

A steamer, on her passage from Ireland to Scotland, was perceived by her commander during the night, and in a smooth sea, to be going with much greater than ordinary velocity through the water. The engineer was not at his post; the Captain inquired of the fireman how it was that the engines were going so fast: the man said 'He could not tell, for he had very little steam and had been firing hard nevertheless'. The Captain began to look about him and, approaching the chimney where the exposed safety valves were fixed, he perceived a passenger fast asleep with the greater part of the weight of his body resting on the flat, cheese-shaped, weights of the safety valve. This man had contrived, with some luggage, to make his bed there for warmth. On arousing and turning him off, the valve rose, the steam escaped with a roar which denoted its having attained a very elevated pressure.

There was *no mercurial gauge* to indicate the pressure of the steam to the fireman who was accustomed to keep it as near as he could to the blowing-off point: and not having heard it escape, he 'fired-up' believing his steam to be low; and he was too ignorant to ascertain the fact, though the increased speed of the engines should have informed him that something unusual had occurred.

It is mentioned by several of our correspondents that engine men, firemen and even masters have frequently been caught *sitting*, or even *standing*, on the safety valves, or hanging weights and resting their bodies on the levers in order to raise the pressure of the steam at the moment of starting.

The report goes on to say that it was also the practice to stow

* She died of T.B. at the age of twenty-seven. What she actually did was more intelligent and much more seamanlike than one would infer from the popular stories and pictures.

surplus *bunker coal* on top of the safety valve. The steamship *Hercules* blew up from this cause. Altogether, it is rather remarkable that only seventy-seven lives were lost from boiler explosions in British steamers during the period under review.

The record of the railways was about as bad as that of the steamships and for much the same causes. There was a succession of very serious accidents extending over a period of seventy or eighty years. About the last of these occurred in 1909. A locomotive boiler blew up although the pressure gauge appeared to be showing zero pressure. It turned out that a workman had assembled the safety valve the wrong way round, so that it was incapable of blowing off at all. The gauge appeared to show no pressure for the simple reason that the needle had gone right round its full travel and was pressing against the wrong side of the stop pin. Three people were killed and three more badly injured.

In these latter days the number of boiler explosions has greatly diminished. This is partly because the manufacture and maintenance of steam boilers is now closely controlled by law and by the insurance companies, but perhaps more because the number of steam engines in service is now quite small and those that do exist are nearly all large plants, such as power stations, which are presumably run by competent people.

But – when is a boiler not a boiler? This is quite an interesting legal question. There exist in industry a large number of pressure vessels of one kind or another which are used in various manufacturing processes. Many of these vessels are of more complicated and less conventional design than traditional boilers and they may be less obviously dangerous. In general the control over their manufacture and use is less strict than for ordinary boilers. However, many of these vessels are heated by process steam or by hot oil under pressure, so that the consequences of fracture may be nearly as bad. It is well to bear in mind that the fatigue limit for the weld metal in mild steel structures exposed to wet steam may be as low as $\pm 2,000$ p.s.i.

In one instance in which I was concerned, two large rotating drums, used for making plastic-coated paper, had been converted from low-pressure oil heating to steam heating – using process steam at a higher pressure. To make certain, the insurance company's inspector had insisted that the drums be 'strengthened' internally by connecting the flat end-plates to the cylindrical surface by means of a number of large triangular gussets, or brackets, cut from mild steel plate and welded in place.

Both drums burst in service after being used for a short time with steam heating. From the drawings I calculated that, in the two drums, there were forty-eight individual places at which failure should have taken place. In fact this was a pessimistic estimate; failure actually occurred at only forty-seven places. By the grace of God, nobody was killed or seriously injured: but it was naughty of the insurance company's inspector, who, I expect, was a diligent and well-meaning little man.

Another case was more tragic. A firm of chemical engineering contractors had bought in from elsewhere a mixing vessel which they installed as part of a plant they were constructing for a customer. Since this mixing vessel was intended to be heated by oil under pressure, the pressurized heating jacket had been subjected to a 'proof test' with cold water. It had withstood a pressure of 65 p.s.i. without obvious damage before it was installed. However, when the plant was commissioned and the jacket was filled with very hot oil at only about 23 p.s.i., the jacket burst after a few hours of service, spraying a man with oil at 280° C, from the effects of which he died a few days afterwards.

According to the report of the official inspector, the accident could only have happened as a consequence of gross mismanagement by my clients, the firm of chemical engineers. As a result these people had become involved in very elaborate and expensive litigation in the High Court.

In fact the official report of the accident was based on faulty observation of the broken remains and was quite misleading. The vessel had burst, not because it was mishandled by my clients, but because it was of incompetent design and manufacture. Although the technical cause of the accident was, in reality, of a slightly subtle nature, both my clients and also the people who actually made the vessel had assumed the design of such a thing to be a trivial problem. In fact the vessel was never really 'designed' at all in any sophisticated sense but was simply put together 'by eye' in a back-street welding shop.

What actually happened was that, during the 'proof loading' the vital welds which held the pressurized heating jacket together were considerably distorted – although nobody noticed it at the time. In reality these welds were so near to failure that a few reversals of stress, resulting from a much lower pressure in the jacket, sufficed to cause a fatigue failure, with disastrous consequences. This possibility should have been spotted by a competent trained engineer. In law, and perhaps in equity, the major blame lay with the people who had made the vessel; but I cannot help thinking that the danger should have been foreseen by a competent firm of chemical *engineers*. When I went to see these people the managing director took me out to lunch. By way of making conversation I said 'How many graduate engineers do you have in your organization, Mr —?'

'None, thank God!'

On cutting holes in things

Although it is generally rash to cut holes in an existing structure some people seem unable to resist the temptation to do so. A case in point occurred with the Master aircraft. This aeroplane was built as an advanced trainer for the R.A.F. just before the war. It had some of the performance, and many of the handling qualities, of the Hurricane and the Spitfire. In the emergency of 1940 some of the Masters were converted into operational fighters by installing six machine guns in the wings. The original trainer version of the machine had wire-operated control surfaces which, though they were perfectly satisfactory, gave a slightly 'softer' response than those of a real fighter. Somebody therefore decided to change over from wire to rod control linkages in the fighter version of the Master. To make room for the rods which operated the rudder and elevator, suitable slots were cut in the rear bulkhead of the fuselage.

Before long we were faced with a series of three fatal accidents. In each case the tail had come off in flight. When we got the fuselage on the test-frame we found that its strength had been reduced to only 45 per cent of the fully factored load. The moral is, I suppose, to leave well alone.

A much better-known accident of this type, in which a great many lives were lost, occurred with the troopship *Birkenhead*. This iron steamship had started life as a warship in 1846 with adequate strength and well supplied with continuous water-tight bulkheads. When she was converted into a troopship, however, the War Office insisted that very large openings should be cut in the transverse water-tight bulkheads* so as to give more light and air and more apparent space for the troops.

In 1852 the *Birkenhead* was dispatched to India, by way of the Cape, with 648 persons on board, including twenty women and children. By an error of pilotage, the ship struck an isolated rock about four miles off the South African coast. The vessel was badly holed forward, and, since the bulkheads had been cut away, all the troop-decks in the forward part of the ship were flooded so quickly that many of the troops were drowned as they lay in their hammocks (the time being 2 a.m.).

Under the weight of the incoming water the flooded fore-part of the ship broke off and sank almost immediately, leaving the survivors crowded into the after-part, which sank more slowly. It was dark, the sea was full of sharks and the life-boats were inadequate. The troops behaved with great courage and discipline, smartly fallen-in on the after-deck, while the women and children were sent ashore in such boats as there were. All of the women and children were saved but only 173 men survived: the rest were drowned or eaten by sharks.

The most obvious effect of cutting holes in the bulkheads was, of course, that the various compartments in the ship flooded very rapidly, and this was undoubtedly the prime cause of the ship's loss. Fewer lives might have been lost, however, if the ship had not broken in two, and this must be attributed, at least in part, to the weakening of the hull as a whole by cutting away the bulkheads on which its strength depended.

The loss of the *Birkenhead* immediately became famous as an example of discipline and heroism – and deservedly so. When the news reached Berlin, the King of Prussia ordered the story to be read aloud to all the units of his army, specially paraded for the purpose. But perhaps it would have been better still if he had

* Except, of course, the engine-room bulkheads.

instructed his War Office not to interfere with the structure of ships, a matter which soldiers do not always understand.

According to Mr K. C. Barnaby, a distinguished naval architect, the idea that open space was more important than safety in troopships lasted for many years. He says that, as late as 1882, shipowners were complaining that, when they fitted additional bulkheads as urged by the Admiralty, the trooping authorities would not accept the ships on the ground that the spaces between the bulkheads were too small.*

On being overweight

Almost every structure has a tendency to turn out heavier than its designer intended. This is partly due to over-optimistic estimating in the weights office, but it is also due to a tendency on the part of almost everybody to 'play safe' by making each part just that much thicker and heavier than is really necessary. In many people's eyes this is a sort of virtue – a sign of honesty and integrity – and we talk of things being 'heavily built' as a term of praise, while 'lightly built' is almost synonymous with 'flimsy' or 'shoddy'.

Sometimes this does not matter, but there are cases where it matters very much indeed. With aircraft the weight is tending to increase all the time, from the drawing-board onwards. Extra weight naturally restricts the fuel capacity or the payload of the aeroplane, but, besides this increase in gross weight, the centre of gravity of an aeroplane somehow always manages to work its way too far aft. In other words the weight of the tail tends to increase out of proportion to that of the rest of the machine. This can be a serious matter. If the C.G. gets too far aft, the aircraft will acquire dangerous flying characteristics. It may have a tendency to go into a spin from which it is unable to recover. For this reason a surprising number of aircraft – including some very famous ones – have gone around all their lives carrying massive lead weights permanently bolted into their noses; this is necessary in order to keep the C.G. in a tolerably safe position. It need hardly be said that this is a bad thing.

*K. C. Barnaby, *Some Ship Disasters and their Causes* (Hutchinson, 1968).

The effects of overweight are just as bad, perhaps worse, with ships. Not only do all ship hulls tend to be overweight absolutely, but the C.G. tends, in this case, to creep, not backwards but upwards – ineluctably upwards. Now the stability of a ship, that is, her tendency to float right-side up, instead of upside-down or on her side, depends upon something called her 'metacentric height'. This is the vertical distance between a mystic but important point called the 'metacentre' and her centre of gravity. For excellent reasons the metacentric height of even a large ship is likely to be quite a small distance – in fact in the region of one or two feet, perhaps less. Thus the position of the C.G. has only to rise by a matter of a few inches to reduce the metacentric height by a very significant fraction which may well imperil the safety of the ship. Various ships have capsized on launching for this reason, and no doubt the yard foremen, or whoever were responsible for the extra top-weight, considered that they were in no way to blame.

We mentioned the loss of H.M.S. *Captain* in Chapter 11. The whole story of the *Captain* was intensely political and controversial at the time; I suppose few accidents can have had such far-reaching historical consequences. The *Captain* represented one turning point in the evolution of the steam battleship and perhaps in the modern concept of world power. The Admiralty have often been criticized by historians who know very little about ships for their slowness in changing from sail to steam. These are sometimes just the historians who are most critical of 'imperialist expansion' and so forth.

It has to be borne in mind that, until comparatively recently, the unreliable engines, the high coal consumption and the short range of steam warships made them dependent upon bases and coaling stations and 'colonies' as soon as they ventured beyond home waters. The exercise of world power by steam navies is a very different sort of thing from the strategy and logistics of eighteenth-century sailing fleets. It was basically for such reasons that the British Admiralty insisted upon the retention of full sail power, in addition to engines, in most of their battleships almost to within living memory.

The technical difficulty of combining sail with steam propulsion lay less in the nature of engines and sails than in the developments which took place during the nineteenth century in guns and armour. Turret guns require a wide angle of fire, besides being very heavy. The necessary protective armour was even heavier. To combine the required fields of fire, and also adequate stability, with full sail propulsion constituted a very difficult problem in naval architecture. In the 1860s the Admiralty were understandably inclined to proceed cautiously. If they had been allowed to continue to do so, all might have been well and history might have been considerably different.

This applecart was upset by a certain Captain Cowper Coles. Coles was one of those clever men with an exceptional talent for controversy and publicity. Having invented a new sort of gun-turret, he set himself to persuade the Admiralty to build a battle-ship around it with full sailing rig and therefore unlimited range. Coles managed to involve, not only the Admiralty, but also both Houses of Parliament, the Royal Family, the Editor of *The Times* and practically the whole of the Establishment in what became one of the greatest publicity exercises of its kind.

Tiring eventually of being called 'reactionary' by half the newspapers and more than half the politicians in the country, the Admiralty gave way. They did what they had never done before, and will certainly never do again; they allowed a serving naval officer with no qualifications in naval architecture to design his own private battleship and have her built at the public expense.

The ship was built by Lairds at Birkenhead as Coles's responsibility and with none of the usual checks on design. She was, moreover, built in a blaze of vituperation and controversy. For much of the time Coles himself was ill and unable to leave his home in the Isle of Wight to attend the yard. As a result of all this muddling, the ship ended up about 15 per cent overweight. If this had not been the case it is at least possible that the ship would have been a success and comparatively safe.

As it was, the *Captain* was much too deep in the water and her C.G. was much too high up. Subsequent calculations showed that the ship would capsize if allowed to heel beyond an angle of 21°. However, the ship was commissioned in 1869 with much publicity. She made two deep-water cruises to the great satisfaction of *The*

Times and of the First Lord of the Admiralty, who had his own midshipman son transferred into her. It looked as if the problems of world power, without the encumbrance and potential embarrassment of world bases, were going to continue to be soluble.

On her third voyage, returning from Gibraltar in 1870 in company with the rest of the Channel Fleet, H.M.S. *Captain* suddenly capsized in a rather moderate squall in the Bay of Biscay. 472 lives were lost – more than the total British dead at Trafalgar. Both Cowper Coles himself and the First Lord's son were drowned. Only seventeen men and one officer were saved.

Though not, of course, the sole factor, the loss of the *Captain* had a powerful effect in accelerating the change from sail to steam, or rather on the abolition of the full sailing rigs in big battleships. Whatever the technical consequences, the political ones were extensive. It will be remembered that the Suez Canal, which was opened just before the *Captain* was launched, originally belonged effectively to France. Disraeli bought the Suez Canal shares for the British government in 1874, and the acquisition of a world-wide chain of coaling stations became a political necessity. The whole story of the *Captain* disaster is complicated, but the immediate technical cause was undoubtedly the determination to ensure that the masts and hull of the ship should have really adequate strength – regardless of weight. It was one of many structural accidents in which nothing actually broke, but the causes were just as 'structural' as if they had.

Aeroelasticity – or a reed shaken by the wind

When a fluid, such as air or water, flows past an obstruction, which might be a tree or a rope, eddies of fluid are formed behind it. Quite often, if you observe a reed or a bulrush growing in a fairly slow-moving river, you will see that the eddies in the sliding water are formed first on one side, then on the other, alternately. The result is a rhythmic variation of fluid pressure, from one flank of the obstruction to the other. Such a succession or 'street' of eddies is called a 'Karman strasse', after the aerodynamicist von Karman, who first described it. It is often quite easy to see eddies on the surface of smooth water though eddies in air are invisible unless they are shown up by smoke or dead leaves or some similar indicator. In fact, however, just the same Karman strasse of eddies happens when air blows past a flag or a tree or a wire. The result of these alternate eddies, acting first on one side then the other, is that the flag flaps, the tree sways and the telegraph wires sing and hum in the wind. Thus a sail will flap as soon as the sheet is eased and may very well split itself or injure somebody. I remember seeing a man knocked out by a flogging sheet-block; there is a lot of energy involved. When a big ship is tacking in a breeze, the noise is as loud as gunfire and much more impressive.

If the frequency of the aerodynamic stimulus provided by the eddies happens to coincide with one of the natural periods of vibration of the obstruction, then the amplitude of the movement may increase until something breaks. It is this sort of thing, rather than steady wind pressure, which usually accounts for trees being blown down. In a somewhat more sophisticated way this is also what is rather too apt to happen with aeroplanes and suspension bridges. It can be prevented by making the structure adequately stiff, especially in torsion. As we have already remarked, it is the torsional stiffness requirements which generally govern the design, and the structure weight, of modern aircraft.

Although Telford's Menai suspension bridge was quite badly damaged by wind-induced oscillations not long after it was built, it took about a century for the reality of this danger to register properly with bridge designers. The classic catastrophe was that of the Tacoma Narrows bridge in America in 1940. This bridge, which had a span of 2,800 feet (840 metres), was built without adequate torsional stiffness. As a result it swayed in even a moderate breeze to such an extent that the locals immediately christened it 'Galloping Gertie'. Quite soon after it was built it swayed and wriggled itself into a dramatic collapse in a wind of only 42 m.p.h. Fortunately somebody happened to be present with a film in their cine-camera. The camera worked and the price of the film must have turned out to be a good investment, since it has been shown repeatedly in practically every engineering school in the world ever since (Plate 20).

In consequence modern suspension bridges are built with adequate stiffness, especially torsional stiffness. As in aircraft, the stiffness requirements account for a good proportion of the weight of the bridge. In the case of the Severn road bridge (Plate 12), for instance, the decking is made from a very large steel tube of flattish six-sided section, built up from mild steel plates. During construction this tube was floated out in sections, which were hoisted into place and then welded into a continuous structure.

Engineering design as applied theology

In nearly all accidents we need to distinguish two different levels of causation. The first is the immediate technical or mechanical reason for the accident; the second is the underlying human reason. It is quite true that design is not a very precise business, that unexpected things happen, that genuine mistakes are made and so forth; but much more often the 'real' reason for an accident is preventable human error.

It is rather fashionable at present to assume that error is one of those things for which it is not really fair to blame people, who, after all were 'doing their best' or are the victims of their upbringing and environment, or the social system – and so on and so on. But error shades off into what it is now very unpopular to call 'sin'. In the course of a long professional life spent, or misspent, in the study of the strength of materials and structures I have had cause to examine a lot of accidents, many of them fatal. I have been forced to the conclusion that very few accidents just 'happen' in a morally neutral way. Nine out of ten accidents are caused, not by more or less abstruse technical effects, but by old-fashioned human sin – often verging on plain wickedness.

Of course I do not mean the more gilded and juicy sins like deliberate murder, large-scale fraud or Sex. It is squalid sins like carelessness, idleness, won't-learn-and-don't-need-to-ask, you-can't-tell-me-anything-about-my-job, pride, jealousy and greed that kill people. Though some engineering firms have splendid design teams, far too many firms in this country are technically incompetent – often to a criminal extent. Many of these people have risen from the shop floor, and, out of a mixture of pride and meanness, they intensely resent any suggestion that they should seek proper advice or employ qualified staff.

It is my experience that far more accidents occur every week than ever get into the papers; generally they are caused by lack of proper care and professional competence. I very much doubt if the remedy lies in the imposition of yet more regulations. It seems to me that what is wanted is the creation of more public awareness and a climate of opinion which regards such 'mistakes' as morally culpable. The man who drilled a hole in the wrong place in the wing-spar of a wooden aeroplane, plugged the hole, and said nothing, was acquitted. Presumably the jury thought that the moral blame was negligible.

What is wanted is much more publicity; the difficulty lies in the law of libel. In most cases, if the real causes of an accident are made public, somebody's face will be very red, and it is likely that their business or professional reputation will suffer. Most practising engineers are acutely aware of this and have to keep quiet or risk heavy damages. In my opinion there should be some way round this, for it is in the public interest that accidents and blunders should be publicized.

Though the great majority of structural accidents are sordid back-street affairs which we hear very little about, there are, of course, a certain number of great dramatic accidents which, for a while, monopolize the headlines. Of such a kind were the Tay bridge collapse in 1879, the capsize of the *Captain* in 1870, and the R101 disaster in 1930. These are very often intensely human and intensely political affairs, caused basically by ambition and pride. The sinking of the *Captain* was of this nature: the two men who carried the heaviest moral responsibility paid heavily for their faults, the one with his own life, the other with that of his son. Unfortunately a great many other lives were lost too.

The wreck of the airship R101, which hit the ground and was burnt out at Beauvais in 1930, was basically similar. There is a splendid account of this by Nevil Shute in his book *Slide Rule*. The immediate technical cause of the accident was the tearing of the fabric of the outer envelope; this fabric had apparently been embrittled by improper doping treatment. The real reason for the disaster was, however, pride and jealousy and political ambition.

The Labour government's Air Minister, Lord Thompson, who carried the ultimate responsibility, was burnt to death in the accident, along with his valet and nearly fifty of the crew.

Nevil Shute's account of the events leading up to the accident corresponds extraordinarily closely in character with my own experience of rather comparable circumstances. One can at once recognize a certain atmosphere of Gadarene inevitability about the whole procedure. Under the pressure of pride and jealousy and ambition and political rivalry, attention is concentrated on the day-to-day details. The broad judgements, the generalship of engineering, end by being impossible. The whole thing becomes unstoppable and slides to disaster before one's eyes. Thus are the purposes of Zeus accomplished. People do not become immune from the classical or theological human weaknesses merely because they are operating in a technical situation, and several of these catastrophes have much of the drama and inevitability of Greek tragedy. It may be that some of our text-books ought to be written by people like Aeschylus or Sophocles – these writers were not humanists.

Chapter 16   Efficiency and aesthetics

– or the world we have to live in

'Why don't you have Mr Smith in your Cabinet, Mr President?'
'I don't like his face.'
'But the poor man can't help his face!'
'Anybody over forty can help their face.'

Told of President Lincoln

Once upon a time I used to work in an explosives laboratory. Naturally very thorough precautions were imposed by the authorities against the entry of unauthorized persons, who not only might sell stolen explosives for a large profit, but might equally well blow the whole place up. Thus this establishment was ringed with barbed wire and alarm bells and armed guards and police dogs and with nearly every device that the ingenuity of security officers could think of.

Now many practical explosives are based on nitro-glycerine, which, by itself, is an exceptionally dangerous liquid both to store and to handle. The least undue familiarity, such as shaking the bottle, may cause it to detonate with the most appalling results. Ordinary safe explosives, such as dynamite, contain a large amount of nitro-glycerine which is only rendered safe to handle by the addition of various substances that have been developed over the years by a succession of rather brave scientists, such as Abel and Nobel. Those who have to experiment with straight nitro-glycerine need to take the most fantastic precautions, and the dangers are such that they not infrequently suffer from nervous breakdowns. Not only are nitro-glycerine laboratories physically separated from other buildings by earthen embankments and wide open spaces, but the staff often wear special clothing, including a peculiar kind of boot devised so that they may tread softly and build up no electrical charges, let alone anything so dangerous as a spark.

One week-end some of the local children managed to wriggle

under the security fence and to evade the police and their dogs. Finding themselves in an apparently lonely place, they broke into one of the nitro-glycerine laboratories. There was, however, nothing very much there to interest them, so they upset the various bottles and beakers of nitro-glycerine on to the floor, stole a couple of pairs of special boots and escaped, by the way they had come, undetected from that day to this.

This is a true story; but I rather think that it might also serve as some sort of parable, for it is possible that engineers and planners and bureaucrats and do-gooders and all the company of the avant-garde are like children playing in a shed full of nitro-glycerine – sublimely unaware that they may cause a major explosion. It is all very well to concentrate on 'efficiency' and making things work, and of course, it is necessary to meet material needs – though in fact our material needs are more flexible than we like to think. However, people have subjective needs which are more important and much more likely to lead to social explosions if they are abused or neglected.

So, when I listen to some of my engineering colleagues talking, I sometimes shake in my shoes. It is not only that they regard the aesthetic consequences of their work as of quite minor importance but that they regard concern about it as basically frivolous. Yet I think that the more we increase material prosperity, the more serious in the long run will be the ultimate catastrophe if people cannot find aesthetic satisfaction.

When I was an engineering student I used to escape from my classes, panting for air, and creep guiltily to the local museum. Many a mathematical lecture did I cut, spending the time looking at the pictures in the Glasgow Art Gallery. No doubt pictures in museums do help, but in a way such things are a pathetic necessity, a refuge of desperation, not only from the aridities of analytical lectures but, more important, from the all-pervasive ugliness of towns like Glasgow.

Of course it suits the tidy philistine administrative mind to keep 'art' in separate boxes called museums and theatres, and it is noticeable that the brave new 1984 regimes provide not only pictures in galleries but also music and ballet. But such forms of 'fine art' can only operate occasionally in the ordinary person's life. They may provide an escape, but they are really no substitute for an environment which is satisfying in itself and is continually present. Most of us find some sort of refreshment in the countryside, but we are pretty well resigned to the dreariness of towns and factories and filling stations and airports and most of the things with which we have to spend our day. Possibly fish which have to live permanently in dirty water may get more or less used to it – but human beings who are conditioned in this way ought to rebel.

We 'Compound for sins [we] are inclined to/By damning those [we] have no mind to.' And, as Professor Macneile Dixon once said,

. . . contrast the middle centuries, that unique period in our European annals, with the centuries following upon the Renaissance. How different their respective views of the world, how opposed their systems of belief! Yet in each the doctrines universally held are felt as inevitable, as unassailable. Each age thinks itself in possession of the true and only view possible for sensible man.*

Thus, about the important things, each age has a totally closed mind. Nowadays, being materialists, we are duly horrified that our ancestors were prepared to tolerate physical poverty and to inflict physical pain. But these same ancestors would be just as horrified that we should suffer many millions of people to experience every day the beastliness of London or New York; and that those who work in our Dark Satanic Mills should have to be well paid to put up with noise and ugliness which are largely unnecessary. Even the 'clinical' decor and atmosphere of modern hospitals would seem to them to add a new terror to dying. Therefore many of us seek some kind of relief or consolation in 'Nature' and we escape, when we can, to the country, because we find the countryside more agreeable than towns and roads and factories. Many people indeed believe that Nature is in some way inherently beautiful and, perhaps, in some way inherently 'good'. Taken to the extreme such views lead to something very like Pantheism – to Meredith's *Woods of Westermain*. But it seems to me that, if we can only get rid of our romantic prejudices and

* W. M. Dixon, *The Human Situation* (Penguin, 1958).

really look at all sides of the question we are forced to the view that Nature is just as aesthetically neutral as she is morally neutral. Mountains and lakes and sunsets may be beautiful, but the sea is often menacing and ugly, and, so far as I have experienced them, primeval forests are frequently places of horror. Most of the European landscape is not really 'natural' at all. The kinds of plants and trees which are allowed to grow have been carefully selected and controlled, and many species have been artificially bred to their present forms, just as much so as the domestic animals. The patterns in which the plants are grown, the whole lay-out of fields and woods and hedges and villages – not to mention drainage and land improvement – are the result of human choice and effort.

Before the eighteenth century, when most landscape was much wilder, educated men had a dread of 'Nature', which implied to them not only physical discomfort, but Pan in the raw. To these people it was the towns which were habitable and attractive, the country which was inhospitable and ugly. Today, when we admire the lovely English landscape we are really admiring something which was deliberately created by the civilized and intelligent English eighteenth-century landlords.

If the country has gone up in the aesthetic world, the towns have certainly come down. Nowadays when we deplore English towns and factories we are deploring the product of philistine reformers and engineers and architects and businessmen and the little grey men who sit in council offices and the bigger grey men who sit in Parliament. Of these people's sins, it is not enough to say that they know not what they do; for we do that which is inherent in our natures – as Plato well knew. It is at least arguable that the countryside is more attractive than the town not because the country is more 'natural' but because town and country were made, by and large, by very different kinds of people. But the first thing is to see ugliness for what it is rather than accepting it as part of the natural order of things.

We do that which is inherent within us. In a world which has an unreasonable admiration for reason we are apt to forget that the human mind is rather like an iceberg. The rational part of our minds, of which we are conscious, is quite small, and, like the visible part of the iceberg, it is supported from underneath by the subconscious mind, which is much larger.

At this point I am only too acutely aware that we are reaching a stage in the argument which is the province of artists and philosophers and psychologists and that I am miserably qualified to blunder into regions where the angels of art criticism fear to tread. I can only plead that necessity knows no laws, that the modern man-made world is hideous, that sheer desperation induces me – a naval architect *manqué* – to stick my neck out. I think it is really important that some sort of view of the aesthetics of technology and engineering and structures should be put forward to engineers and technologists by one of themselves, however inadequate that view may be. For what follows I commit myself to Athena and to Apollo – by their grace may somebody more competent than myself be provoked into doing the job better.

Let us begin by looking at the human reception process in aesthetics; that is to say, why we react as we do to some inanimate object. Within the subconscious mind there lies an enormous store of potential reactions and 'forgotten' memories. This material is partly inherited genetically from a remote past (Jung's 'collective unconscious') and partly acquired by the individual himself during the course of his own life, mainly from apparently forgotten experiences – sometimes unpleasant ones. Now our physical senses – sight, hearing, smell and touch – continually pass to our brains far more information about our surroundings than our conscious mind can accept or be aware of. But the subconscious is monitoring this information all the time and it is full of receptors and trip-wires which are liable to be influenced by every shape and every line, every colour and every smell, every texture and every sound. We may be totally unconscious of this, but it is happening all the same and it is building up subjective emotional experiences within us – be the effects good or bad.

This sort of process may account in some measure for the way in which we are influenced, subjectively, by inanimate objects and especially in the present context by artefacts. Artefacts are made by people and somebody, at some stage, has some sort of choice in the shape and the design.

It is impossible to make any object without making a series of statements in the process. Even a straight line is saying in effect 'Look, I am straight, not crooked.' Even a very simple artefact contains a package of such statements which have been made by people.

Just as there can be no such thing as a totally objective experience, so there can be no such thing as a totally objective statement – one with no emotional connotations of any kind. This is true whether the statement be made in words or music or colour or shape or line or texture or in what engineers call design.

This brings us from what might be called the 'aesthetic reception process' to the 'aesthetic transmission process'. In other words, how do things come to be designed as they are? What is it that the maker or the designer puts into an artefact which causes it to have the aesthetic effects which it does? The short answer is, to a large extent, 'His own character and his own values.'

Thus whatever we make and whatever we do we nearly always leave upon the thing or upon the action the imprint of our personalities, written in a code which can usually only be read at the subconscious level. For instance our voices, our handwriting and our manner of walking are quite characteristic and are usually difficult to disguise or to imitate. But this sort of thing extends much further than these familiar examples. One dark evening I was in a yacht anchored in a remote Scottish loch. Round the corner of the land, three or four miles away, there came another sailing yacht which I had never seen before and of which I had no knowledge. Though it was quite impossible to recognize her name or her crew I said to my wife 'That boat is being sailed by Professor Thom.' And so she was – for the way in which a man sails a ship to windward is quite as individual as his voice or his writing, and, once seen, can hardly be forgotten. In the same way one can often tell which of one's friends is flying a light aircraft, for the manner of flying shows, unmistakably, the imprint of the character. In the field of painting and drawing, even the work of very amateur performers is apt to tell one more about themselves than about their subjects. Again, it requires exceptional skill to imitate really plausibly the work of a particular artist. Naturally there is no sharp line between painting and drawing and technological design, and almost everything that gets made is likely to carry with it *something* of the personality of the maker.

What is true of individuals is also apt to be true of a society, a culture or an age. Archaeologists can usually date artefacts, such as potsherds, within a very few years on 'stylistic' grounds. If you walk around Pompeii and Herculaneum, you will come away with a quite surprisingly powerful sense of what sort of people the inhabitants were. This has little or nothing to do with the technology of things like the plumbing, and it is something which no amount of factual history can convey. So far this sort of pattern recognition has eluded the computer; long may it continue to do so.

Recently, I was drinking canned beer with a much respected colleague. I said – rather unwisely and priggishly, I suppose – 'Really a thing like this beer-can seems to me to epitomize all the dreariness and commercialism that is wrong with technology nowadays.'

My much respected colleague was down on me like a ton of bricks. 'I suppose you want to sell beer in pitchers or wooden barrels or wine-skins or something. What else would you sell beer in in this day and age except tin cans? How stupid and impractical and reactionary can you be?'

But, with respect, my much respected colleague was missing the whole point. It is not what you do but how you do it that matters. Beer containers are not beautiful or ugly because of the material from which they are made, or even because they are mass-produced. Whatever they are made out of they will convey, unavoidably, the values of the people who are responsible for them. We happen to be a society which is unable to make attractive beer-cans. Indeed we are, I fear, an age rather noticeably lacking in inherent grace and charm.

Greek amphorae were beautiful, not because they held wine and were made of clay, but because the Greeks made them. They were, in their day, simply the cheapest containers for wine. If the Greeks had made tin beer-cans perhaps we should now have collections of classical beer-cans in museums, much admired by artists.

I believe that very few artefacts are intrinsically ugly or beautiful simply because of their function*; they are rather mirrors to an age, to a set of values. Rather the same conditions obtained during the eighteenth century as in Ancient Greece – partly no doubt because it was a classical age which consciously modelled itself on the ancient world. Nearly everything the eighteenth-century craftsman touched was elegant. This was not just a matter of the luxury trade; it extended right through society.

Of course this begs the whole question of 'absolute' standards in aesthetics. Are not 'my' values as good as 'yours', however deplorable and uneducated you may consider my taste to be? Well, I for one feel strongly that there *are* absolute standards in aesthetics which change only gradually through the ages. The modern fashion for 'aesthetic democracy' seems to me perverse and nihilistic and based largely on a desire to bash the Establishment. I would take the view that there is a continuing tradition of values in aesthetics – just as there is in ethics. The process is an iterative one, advancing slowly and painfully from age to age and from fashion to fashion, building, like science, on the experience of the past. Otherwise how are civilized values ever to be built up?

Another debatable point is 'Granted that common objects such as Greek amphorae were beautiful in some absolute sense, did the Greeks *realize* that they were beautiful?' I am reminded of a remark in a leading article in *The Times*, which said something like 'Good typography should be like clean glass – one should be able to see through it without being distracted. But if this is to happen then the typography must have that sort of discreet elegance and beauty which draws no attention to itself.' I think this is why we only come to appreciate many common artefacts after they have passed out of common daily use. This does not mean that they are not absolutely and permanently beautiful.

And the eighteenth century invented the Industrial Revolution. I think it is important to point out that many of the fathers of the Industrial Revolution were not philistines but sensitive men of

* *Vide* the recent vogue for collecting chamber-pots. Aristophanes regarded Greek oil-bottles as essentially ridiculous but he never implied that they were ugly: indeed the ones in museums are much admired.

considerable taste. Of such a kind were Matthew Boulton (1728–1809) and Josiah Wedgwood (1730–95). They made a great deal of money, the things they made were beautiful, and these two at least were model employers. No doubt there were black sheep, but the evils of the Industrial Revolution did not lie in the ethic of eighteenth-century culture and classicism but rather in a newly arisen vulgarity and greed which came, I think, from outside this ethic.

Neither mass-production machinery itself nor its products are intrinsically ugly. The very first real mass-production machinery, the well-known block-making equipment installed around 1800 at Portsmouth Dockyard by Sir Marc Brunel, is handsome and satisfying. These machines were not only good-looking but also very effective, for they turned out automatically all the millions of pulley blocks needed by the sailing navy during the Napoleonic Wars and for long afterwards. They saved a vast amount of money in doing so, for blocks are expensive things and a single warship might require 1,500 of them. Some of this machinery can now be seen in the Science Museum (Plate 21), but a good deal of it is still in service at Portsmouth after 180 years, supplying the modern navy's diminished need for blocks. Not only the machinery but the product, the blocks themselves, is solid and handsome; whether you would call a block beautiful is a matter of opinion but they are certainly pleasant to look at.

Sir Marc – father of the great Isambard Kingdom Brunel – was a French royalist émigré, and all accounts agree that he was a charming man. We are told that

> The dear old man had, with a great deal more warmth than belonged to that school, the manner, bearing and address and even the dress of a French gentleman of the ancient régime, for he had kept to a rather antiquated but very becoming costume. I was perfectly charmed with him at our first meeting. What I loved in old Brunel was his expansive taste and his love or ardent sympathy for things he did not understand or had not had time to learn. What I most admired of all was his thorough simplicity and unworldliness of character, his indifference to mere lucre, and his genuine absent-mindedness. Evidently he had lived as if there were no rogues in the world.

No doubt a very impractical sort of character who would find difficulty in getting a job with a modern go-ahead firm. But his machinery is still producing blocks, nearly two hundred years after he made it – and it is beautiful.

The great engineers who worked before and immediately after 1800 between them laid the foundations, not only of British industrial prosperity, but of the modern technological world. Many of these people were men of taste. But by the time Queen Victoria came to the throne public taste was undoubtedly deteriorating: by 1851 it had reached an all-time low. Shrewd observers, like Lord Playfair (1818–98), were, however, already remarking, as early as the time of the Great Exhibition, that British industry was losing its impetus and its creativity. Although it is very widely and commonly believed – indeed taken as axiomatic – that ugliness came in with industrialism as an inescapable consequence of mass-production, I doubt if this view would really stand up to proper historical examination. I think it is more reasonable to suppose that elegance and business enterprise declined more or less hand in hand and as a result of something rather nasty and complacent which emerged from the British character during the Age of Reform.

The passionate protest of the Aesthetic movement in the 1870s and 1880s against the ugliness of pretty well everything failed to have much effect. I think this was less because these people were guyed by Gilbert and Sullivan in *Patience* and in the pages of *Punch* than because the movement was largely an escapist one and attacked the wrong targets. These sons of Mary failed to see that the root cause of all the brazen horrors which they hated so much lay, not in machinery itself, but in attitudes of mind. Like so many aesthetic reformers, they rejected technology instead of joining it. Perhaps if they had been prepared to learn technology and engineering they might have operated from within the system. But this is a laborious discipline which too many Arts people reject as being somehow beneath them. Of course William Morris and his followers studied and practised various small-scale technical crafts; but what was needed was to come to terms with real mass-production machinery and with the economic problems of a high-production society.

On efficiency and functionalism

But when his disciples saw it they had indignation, saying, 'To what purpose is this waste? For this ointment might have been sold for much, and given to the poor.'

Matthew 26.8–9

Although we may justly accuse modern engineers of philistinism, nearly all of them do cling to certain very important values which are unfashionable and unpopular in a permissive age. The chief of these are objectivity and responsibility. Engineers have to deal, not only with people and all their quirks and weaknesses, but also with physical facts. One can sometimes argue with people, and it is not difficult to deceive them; but it is of no use to argue with a physical fact. One cannot bully it or bribe it or legislate against it or pretend that the truth is something different or that the thing never happened at all. Laymen and politicians may create what fantasies they choose, but, for the engineers, 'It is their care that the gear engages; it is their care that the switches lock.' Essentially, these people's stuff must work, and go on working, safely and economically. It may be the engineer's job to point out that the emperor has no clothes on, but however embarrassing this may be, we clearly need more, not less, of this kind of realism.

In the pursuit of their objective profession, engineers have developed a number of concepts which are useful as aids to realism. One of these is 'efficiency'. Thus it is very helpful to know what fraction of the expensive energy which is fed into an engine as fuel emerges as useful power. This can be expressed as a simple ratio or percentage, and it tells us a most important fact about one aspect of the working of the engine. Again, it is valuable to be able to compare the weights and costs and load-carrying capacities of various kinds of structures. As we saw in Chapter 14, there are various numerical ways of doing this.

But the concept of efficiency is so useful, and sometimes so economically powerful, that there is a danger of being carried away by it. If we try to apply the idea of efficiency to the *totality* of a situation, then we are usually presuming to a wisdom, to a knowledge of *all* the facts, which is most unlikely in mortal man. We may fairly talk of the efficiency of an engine in terms of fuel consumption and power output: if we talk of the 'efficiency of the engine' – *tout court* – we are being hubristic. We take no account, for instance, of the noise and smell which the engine makes. Or whether the man who has to start it is likely to have heart failure. Or how much pleasure anybody derives from its appearance.

Even if we know all the relevant facts about any technological situation, which is impossible, we could not weight them or quantify them, for many of them are incommensurable. Not long ago there was a great to-do about the proposal to build a vast airport on the Essex coast. This was a project to put down a hideous mass of concrete and sheds and machinery upon the wet, ribbed sands of the Thames Estuary, where the gulls paddle and wheel and squawk. The politicians and the administrators and the economists and the engineers were full of facts and figures about the need for another airport. But it is impossible by any numerical criterion to compare the claims of the planners and the economists with the rights of the gulls and with the beauty of the wet sands. For myself, I am passionately on the side of the gulls, and it gives me immense pleasure to think of all those miles of wet sand and mud, which, I am glad to say, is quite useless and unproductive. So far, the gulls and the sands seem to be winning.

I suppose that it is possible to measure the 'efficiency' of an airport in terms of how many aircraft and passengers it can handle in relation to the capital costs and running costs, and these figures have some practical value, even if they bear no relation in this world to seagulls and wet sand. But for many things the concept of efficiency is simply irrelevant. It is meaningless to talk about the 'efficiency' of a piece of furniture or of a cathedral. All the same, engineers cling to the idea that it 'ought' to be possible in some way to measure the 'efficiency' of practically everything. But this is nonsense.

'Very well,' says the engineer, 'but things must be functional; the beauty of technology lies in its functionalism.' If by this he means that things must work and do their job properly, then he is merely stating the obvious. But when we come to apply functionalism as an aesthetic criterion we are apt to get into some very deep water. There are certain structures, such as bridges, where the structural function is simple and obvious and proclaims itself as such. Many of these are beautiful, but some of them are not. There are also a certain number of very expensive artefacts which are certainly good-looking, such as Concorde and the Rolls Royce car. But are we sure that we are not admiring perfection of workmanship, purchased almost regardless of cost? Ought we not to take cost into account in assessing functionalism?

Now a Ford car can be bought for something like a tenth of the cost of a Rolls, and in the real world, where things have to be paid for, many people would regard the Ford as more 'functional' than the Rolls. But the external appearance of the Ford bears little relation to its mechanical workings; what we see is more or less a tin box put round the machinery by the bodymakers and the stylists. The mechanical, that is to say, the functional, parts of any modern mass-produced car are not attractive, being made largely from bits of wire and bent metal which we find it difficult to admire, however useful they may be.

In the same sort of way, most electrical devices such as wireless sets are hideous in their naked wiry state, and we are constrained to hide them inside black or grey or walnut boxes. On the whole it may be fair to say that, as modern technology gets more and more functional, we can less and less bear to look at it.

But have we not good precedents in Nature? The outside of a person or an animal may be very beautiful; the inside is generally repulsive. Our admiration of Nature is highly selective. We admire certain stages of growth (lambs but not foetuses); we are generally horrified by decay and all those worms. But decay is just as necessary and just as functional as growth.

With regard to this question of functionalism and 'efficiency' Nature seems to have a sense of humour, or perhaps just a sense of proportion. She will construct the stem of a plant, for instance, with the uttermost regard for metabolic economy; the thing is a miracle of structural efficiency. Having done this, she will put a great big flower on top – for fun, as far as one can see. In the same way, peacocks have tails and girls have hair which cannot be considered strictly functional. If it be urged by some dreary person that these things are done to encourage reproduction, this is only putting the argument back by one notch. For why should these ornaments be attractive, sexually or otherwise?

Although it is practically an article of religion with many engineers to believe in a close connection between functional 'efficiency' and appearance, I am, myself, sceptical. Of course, the grossly ineffectual will, and should, offend the eye, but I doubt if the refinements of technical performance really improve appearance very much. Very often it is the other way round; the pursuit of the last ounce of performance results in a boring appearance, as one can see in modern yachts. For myself, I stick to the belief that what one gets aesthetically from an artefact is some combination of the personality of the maker with the accepted values of his age. If you walk down any street with your eyes and your mind open you can form your own judgement on both.

'Science' has been attacked on almost every conceivable ground ever since the Renaissance; most of these attacks were more or less rubbish. But it is always strange to me that what seems the real argument against science is seldom raised, at least in a direct form. This is that science has subtly warped our system of values by teaching us to judge on grounds which are excessively functional. The modern man asks 'What is this man or this thing *for*?' rather than 'What *is* this man or this thing?' Herein, no doubt, lie the causes of many of our modern sicknesses. The aesthetic judgement seeks, however inadequately, to answer the broader and the more important question. Too often nowadays our subjective judgement clashes with our scientific (or banausic) judgement. But we sweep the aesthetic judgement under the carpet at our peril.

Naturally there is nothing in all this to prevent a beautiful object from also being an efficient one. The point I am making is that the two qualities are what the mathematician would call 'independent variables'. I am reminded of the Irish yachtsman's remark: 'An ugly ship is no more attractive than an ugly woman – however fast she may be'.

On formalism and stresses

Modern art and architecture make a great parade about their freedom from traditional forms and conventions – which is possibly why they have achieved so little. Yet formality in design or in manners is not a handicap; such conventions protect the weak and aid the strong. All the loveliest ships have been designed within a stylistic tradition, and I cannot imagine that their designers felt cramped by it. The Greek dramatists wrote within a strict set of rules, but it would be as absurd to think that the *Antigone* is limited by the dramatic unities as to suppose that Jane Austen would somehow have been able to produce greater masterpieces if she had felt free to make use of bad language and overt sex. Of course, fully to appreciate formal achievement it is necessary to have some knowledge of the rules. This applies just as much to the appreciation of cathedrals and bridges and ships as it does to watching cricket. This provides one good reason for knowing something about the principles of engineering as well as the history of art and architecture.

When Ictinus designed the Parthenon in 446 B.C. he worked within the well-established Doric order of architecture. The Parthenon, the Temple of the Maiden, is indisputably one of the most beautiful buildings in the world – possibly the greatest of all artefacts. Although it is dedicated to the divine Athena it is, to me, the supreme statement of humanism – of what the scientist Humphry Davy called the 'brilliant but delusive dreams concerning the infinite improvability of man'. Furthermore, it was built at the very peak of Athenian power and glory and it speaks of the city of the Maiden,

> *Rich and renowned and violet-crowned,*
> *Athens the envied of nations.*

Nemesis, of course, lay just around the corner, very much as it did in 1914. When it was new, in all its white marble, red and blue paint and gilded bronze, the Parthenon might have been just a little vulgar, like some of Kipling. But is not great art always a little vulgar? If the Parthenon is a monument of humanism, some of the earlier Doric temples, say those of Paestum, seem to me to express a moving religious feeling. Contrariwise, the Temple of Hephaistos in Athens, I think, conveys very little – except a faint whiff of commercialism, like Birmingham Town Hall. Yet all these different effects were produced by architects working within a single rigid language.

As with all great art there are many ways of interpreting the Parthenon. What is beyond argument is the magnitude of the achievement. But how did Ictinus do it, working as he did within a strict stylistic convention? Naturally, only one man really knew the answers and that was Ictinus himself; he wrote a book about it, which is now lost. We can, however, make some rather crude analytical observations.

In the traditional, formal steam yacht, grace and majesty are produced by extreme delicacy and subtlety and harmony in the curves of the hull and the sweep of the sheer – by the *exact* and loving placing of masts and funnel and superstructure (Plate 22). *Mutatis mutandis*, this is like the exact and loving placing of words in writing. Ship design differs from the creation of poetry only in its numerate content. So again in Doric architecture, it is the loving attention to detail which is important. Although it appears to be rectangular, there is scarcely a straight line in the Parthenon, and few lines are truly parallel. The seventy-two columns are inclined towards each other in such a way that, if produced, they would all meet at a single point, about five miles up in the sky. The eye, which expects a simple box-like structure, is deceived and enchanted by subtlety after subtlety. Like a clever woman, the Parthenon influences us and bewitches us, though we are scarcely aware of how it is done – or even that it is happening at all (Plate 23).

But what has all this to do with stresses? In one sense a great deal; in another very little. As long ago as the seventeenth century, Fénelon observed that classical architecture owes its effects to the fact that it appears to be heavier than it really is, Gothic to the fact that it appears to be lighter than is really the case. In this respect there appears to be no aesthetic pay-off from honest functionalism – from appearing to be just as heavy as you really are.

The classical orders, especially Doric, appear almost to stagger under the burden of their own weight. In fact there is really very little load in most of the columns, but the swelling or 'entasis' which is given to them provides a sort of Poisson's ratio effect to convince us that they are bulging under the compressive stress. This bulging effect is carried still further by the swelling, cushion-like capitals or 'echinoi' which transmit the compressive load from the lintels to the heads of the columns. The effect of weight is enhanced still further by the excessive depth of the architraves.

Although classical architecture operates on the emotions, at least in part through a subjective sense of stress, its beauty has little or nothing to do with modern ideas of structural efficiency, in the sense of the one-hoss shay. All these buildings were, in fact, thoroughly inefficient. The compressive stresses were absurdly low, while the tensile stresses in the lintels were far too high, often dangerously so (Chapter 9). The roofs of classical buildings, as we have seen, can only be described as a structural mess. But there is nothing wrong with most of these buildings aesthetically.

When we come to consider Gothic architecture, the compressive stresses in the masonry are, as a rule, a good deal higher than they are in classical buildings, and the structure as a whole is generally more stable, in spite of its airy-fairy appearance. The effect of lightness is, however, achieved, in part, by the use of pointed arches, which, again, are 'inefficient'. These Gothic structures are, to the modern functional mind, excessively complicated. The real heroes of Gothic cathedrals seem to be the statues, whose weight, perched on pinnacles and flying buttresses, keeps the thrust lines stable (Chapter 9).

Structurally 'inefficient' as ancient buildings may have been, it does seem that the eye requires some subjective sense of stress if it is to find satisfaction in looking at a structure. In many modern buildings the load-bearing structure, which is often of reinforced concrete, is hidden away inside the building. All that the external observer can see is a curtain wall or 'cladding' of thin brick or glass which is obviously inadequate to carry any load at all. I do not think I am alone in finding these buildings unsatisfactory to look at and often downright ugly.

But supposing that we had some kind of structure whose means of support were clearly visible, and which was also highly 'efficient' in the modern manner, what might we expect it to look like? Clearly this is a subject about which one could argue for a long time. However, if we may judge from the structures which are employed for landing on the Moon – in which weight has been saved regardless of cost, the ultimate in one-hoss shays – the answer seems likely to be 'Hideously ugly'.

On skiamorphs, fakes and ornament

The earliest surviving buildings of consequence in Greece are Mycenaean and date perhaps from some time before 1,500 B.C. These buildings were made of stone and seem to have been deliberately and intelligently designed as structures suited to the characteristics of that material. The Mycenaeans were well aware, for instance, of the danger of excessive tensile stresses in stone lintels, and they made adequate provision to relieve the bending loads on their stone beams, as one can see in the Lion Gate at Mycenae (Plate 24). To this extent, at least, Mycenaean architecture can be described as 'structurally functional'.

When the Mycenaean civilization collapsed, around 1,400 B.C., Greece seems to have reverted to a dark and illiterate age, from which no buildings of any importance survive. No doubt people lived and worshipped in wooden huts of one kind or another. When formal architecture began to revive in early Archaic times, perhaps about 800 B.C., the early temples were built of wood, like the New England churches.

Naturally, none of the original wooden temples has survived. However, the transition from wood to stone construction seems to have been a piecemeal process; as timber became scarce, decayed wooden members were replaced by stone copies. Pausanias speaks of a temple still existing at Olympia in the second century A.D. in which some of the wooden columns still remained, mixed with more recent stone ones.

Doric architecture is thus 'trabeate' or beam architecture, based on wooden construction; and even when temples came to be built, *de novo*, entirely of stone, architects still stuck to the forms and proportions which were suited to timber. Not only did classical architects of the sophisticated fifth century use weak stone beams in the place of wooden lintels; they went to the trouble of copying in marble all sorts of irrelevant constructional details, such as the ends of the wooden pegs which had once held the wooden buildings together.

The result 'ought' to have been ridiculous, but it was not; it was gloriously and triumphantly successful and has served as a model for the civilized world, on and off, for two thousand years. Survivals of this sort are known as 'skiamorphs' (shadow shapes), and in one form or another they are very common in technology. A modern instance is the survival of timber graining on the surfaces of plastic mouldings and furniture.

Contrary to the whole ethic of the functionalist school of engineering aesthetic thought, skiamorphs are not necessarily shoddy or vulgar. Nowadays, of course, they very often are; but surely this is because of our own faulty execution, not because there is something inherently wrong with the idea.

The development of the Watson steam yacht is a splendid example of a successful skiamorph. The classical form for large steam yachts was evolved in latish Victorian times by the greatest of all yacht designers, G. L. Watson (who had for his epitaph 'Justice to the line and equity to the plummet'). For his fully powered vessels Watson retained, not only the graceful 'clipper' bow of the sailing ship, but also the now functionless bowsprit. The result is one of the most beautiful ship conventions which has ever been developed (Plate 22).

If all this be so, what are we to think about 'honesty' in design? Honesty compels me to say 'Not much'. If skiamorphs are permissible in Greek temples and steam yachts, what are we to think of the total 'fake'? Is there any reason why we should not dress up suspension bridges as medieval castles, motor cars to look like stage coaches, or yew-trees to look like peacocks?

Personally I am rather in favour of it. After all, the results could hardly look worse or more depressing than the results of modern functionalism, and they might be a lot more fun. What is wrong with eighteenth-century 'Gothick' buildings? The best of them are tremendous fun and perfectly lovely. Horace Walpole was no fool, and the Pavilion at Brighton is a delight.

There are those who moan about 'meaningless ornament'; but the phrase is surely an oxymoron, for no ornament can be 'meaningless' – even if it means something pretty frightful. If the critic wants to imply 'ornament which is unsuitable or unrelated to its substrate', that is fair enough; but *all* ornament must have some

effect. It seems to me that what we want is more, not less ornament. The truth seems to be that we are frightened to express
ourselves in ornament. We don't know how to handle it, and fear
that we may expose the nakedness of our mean little souls.
Medieval masons did not have that kind of inhibition, and they
were probably psychologically healthier in consequence.

Is it not fair to ask the technologist, not only to provide
artefacts which work, but also to provide beauty, even in the
common street, and, above all, to provide *fun*? Otherwise technology will die of boredom. Let us have lots of ornament. Let
there be figure-heads on ships, gilded rosettes on the spandrels of
bridges, statues on buildings, crinolines on women, and, everywhere, lots and lots of flags. Since we have created a whole
menagerie full of new artefacts, motor cars, refrigerators, wireless
sets and the Lord knows what, let us sit down and think what fun
we can have in devising new kinds of decorations for them.

Afterthought (1980) Since writing this chapter I have come across
a saying of Henry James, 'What is character but the determination of incident? What is incident but the illustration of character?' It is a pity that Henry James was so contemptuous of
technology; he might have had so much to contribute.

Appendix 1   Handbooks and formulae

Over the last 150 years the theoretical elasticians have analysed the stresses and deflections in structures of almost every conceivable shape when subjected to all sorts and conditions of loads. This is all very well, but usually the results, in the raw form as published by these people, are too mathematical and too complicated to be of much direct use to ordinary human beings who are in a hurry to design something fairly simple.

Fortunately a great deal of this information has been reduced to a set of standard cases or examples the answers to which can be expressed in the form of quite simple formulae. Formulae of this sort, covering almost any possible structural contingency, are to be found in handbooks, notably R. J. Roark's *Formulas for Stress and Strain* (McGraw-Hill). These formulae can be used by people like you and me equipped with little more than common sense, a knowledge of elementary algebra and the contents of Chapter 3. A few of these formulae are given in Appendixes 2 and 3 which follow.

Used with caution, such formulae really are very useful indeed, and indeed they form the professional stock-in-trade of most engineering designers and draughtsmen. There is not the slightest need to be ashamed of using them; in fact we all do. But they *must* be used with caution.

1. Make sure that you really understand what the formula is about.

2. Make sure that it really does apply to your particular case.

3. Remember, remember, remember, that these formulae take no account of stress concentrations or other special local conditions.

After this, plug the appropriate loads and dimensions into the formula – making sure that the units are consistent and that the noughts are right. Then do a little elementary arithmetic and out will drop a figure representing a stress or a deflection.

Now look at this figure with a nasty suspicious eye and think if it *looks* and *feels* right. In any case you had better check your arithmetic; are you sure that you haven't dropped a two?

Naturally neither mathematics nor handbook formulae will 'design' a structure for us. We have to do the designing ourselves in the light of such experience and wisdom and intuition as we may possess; when we have done this the calculations will analyse the design for us and tell us, at least approximately, what stresses and deflections to expect.

In practice therefore design procedure often runs something like this. First, one determines the greatest loads to which the structure may be subjected and the deflections which can be allowed. Both of these are sometimes laid down by existing rules and regulations, but, where this is not the case, they may not be particularly easy to determine. This sort of thing calls for judgement, and in case of doubt it is clearly better to err on the conservative side, although, as we have seen, it is quite possible to go too far and incur danger from too much weight in the wrong places.

When the loading conditions have been determined we can sketch out, to scale, a rough design – designers often use pads of squared paper for their preliminary sketches – and we can then apply the appropriate formulae to see what the stresses and deflections are going to look like. At the first shot these will probably be too high or too low, and so we go on modifying our sketches until they seem about right.

When all this has been done, 'proper' drawings may have to be made from which the thing can be manufactured. Formal engineering drawings are very necessary when components have to be made by the usual industrial procedures, but they are troublesome to make and may not be needed for simple jobs or amateur work. For anything of a commercial and potentially dangerous nature, however, it is my experience that a firm can look remarkably silly in a court of law if the only 'drawing' they can produce is a sketch on the back of an envelope.

When you have got as far as a working drawing, if the structure you propose to have made is an important one, the next thing to do, and a very right and proper thing, is to worry about it like blazes. When I was concerned with the introduction of plastic components into aircraft I used to lie awake night after night worrying about them, and I attribute the fact that none of these components ever gave trouble almost entirely to the beneficent effects of worry. It is confidence that causes accidents and worry which prevents them. So go over your sums not once or twice but again and again and again.

Appendix 2   Beam theory

The basic formula for the stress s at a point P distant y from the neutral axis of a beam is

$$\frac{s}{y} = \frac{M}{I} = \frac{E}{r}$$

so

$$s = \frac{My}{I}$$

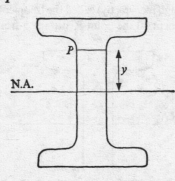

Figure 1.

where s = tensile or compressive stress (p.s.i., N/m² etc.)

         y = distance from neutral axis (inches or metres)

         I = second moment of area of cross-section about the neutral axis (inches⁴ or metres⁴)

         E = Young's modulus (p.s.i., N/m² etc.)

         r = radius of curvatures of the beam at the section under consideration due to the elastic deflections set up by the bending moment M (M in inch-pounds, Newton-metres etc.).

Position of neutral axis

The 'neutral axis' (N.A.) will always pass through the centroid ('centre of gravity') of the cross-section. For symmetrical sections, such as rectangles, tubes, 'I' sections etc., the centroid will be in the 'middle' or centre of symmetry. For other sections, it can be calculated by mathematical methods. For some simple asymmetrical sections (e.g. railway lines) one can determine the centroid accurately enough by balancing a cardboard model of the section on a pin. For more elaborate structures, such as ships' hulls, the position of the neutral axis really will have to be calculated by sheer arithmetic.

'I', the second moment of area of a cross-section

This is often (though incorrectly) called the 'moment of inertia'.

Thus, if an element, at the point P, distant y from the neutral

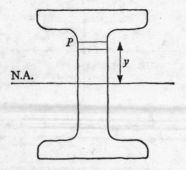

Figure 2.

axis, has a cross-sectional area a, say, then the second moment of area of this element about the neutral axis will be ay^2.

Thus the *total I* or second moment of area of the cross-section is the *sum* of all such elements, i.e.

$$I= \sum_{\text{bottom}}^{\text{top}} ay^2$$

For irregular sections this can be calculated by arithmetic, or there is a version of 'Simpson's Rule' which gives the answer.

For simple symmetrical sections:

For a rectangle about the neutral axis,

$$I = \frac{bd^3}{12}$$

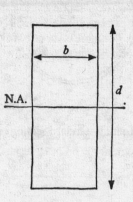

Figure 3.

For a circle about the neutral axis.

$$I = \frac{\pi r^4}{4}$$

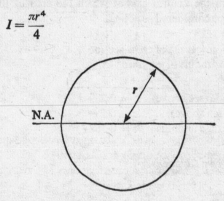

Figure 4.

Thus simple box and H sections as well as hollow tubes can be calculated by subtraction.

For a thin-walled tube of wall thickness t, however,

$$I = \pi r^3 t$$

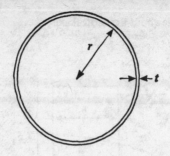

Figure 5.

The *I*s of a great many standard sections can be looked up in reference books.

'Radius of gyration', k

For some purposes, it is useful to know the value of what is called the 'radius of gyration' of a beam section: that is to say, the distance from the neutral axis at which the area of the cross-section may be considered as acting.

i.e. $I = Ak^2$

where A = total area of cross-section

k = 'radius of gyration'

For a rectangle (see above) $k = 0.289\,d$

For a circle (see above) $k = 0.5\,r$

For a thin-walled annulus $k = 0.707\,r$

Some stock beam situations

CANTILEVERS

1. Point load *W* at end
   Condition at distance *x* from end of beam:
   $M = Wx$ Max $M = WL$ at B

   Deflection at *x* is $y = \dfrac{1}{6}\dfrac{W}{EI}(x^3 - 3L^2x + 2L^3)$

Max deflection $y_{max} = \dfrac{1}{3}\dfrac{WL^3}{EI}$ at A

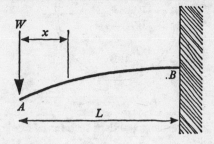

Figure 6.

2. Uniformly distributed load $W = wL$

$M = \dfrac{1}{2}\dfrac{W}{L}x^2$ at x

$M_{max} = \dfrac{1}{2}WL$ at B

Deflection at x is $y = \dfrac{1}{24}\dfrac{W}{EIL}(x^4 - 4L^3x + 3L^4)$

Max deflection at tip

$y_{max} = \dfrac{1}{8}\dfrac{WL^3}{EI}$

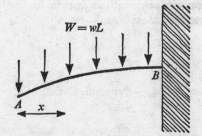

Figure 7.

SIMPLY SUPPORTED BEAMS

3. Simply supported beam with load in centre
   Bending moment M at point x
   (A to B) $M = \frac{1}{2}Wx$
   (B to C) $M = \frac{1}{2}W(L-x)$

$$M_{max} = \frac{WL}{4} \text{ at B}$$

Deflection y at x

(A to B) $$y = \frac{1}{48}\frac{W}{EI}(3L^2x - 4x^3)$$

$$y_{max} = \frac{1}{48}\frac{WL^3}{EI} \text{ at B}$$

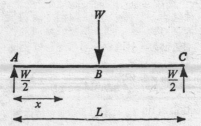

Figure 8.

4. Simply supported beam with single point load not in centre
   Bending moment M at point x

(A to B) $$M = W\frac{b}{L}x$$

(B to C) $$M = W\frac{a}{L}(L-x)$$

$$M_{max} = W\frac{ab}{L} \text{ at B}$$

Max deflection $y = \dfrac{Wab}{27EIL}(a+2b)\sqrt{3a(a+2b)}$

at $x = \sqrt{\tfrac{1}{3}a(a+2b)}$ when $a > b$

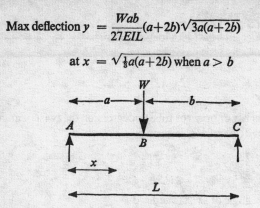

Figure 9.

5. Simply supported beam with uniform load

$$W = wL$$

at point x:

$$M = \tfrac{1}{2}W\left(x - \dfrac{x^2}{L}\right)$$

$$M_{max} = \dfrac{WL}{8} \text{ at mid-point}$$

Max deflection $y_{max} = \dfrac{5}{384}\dfrac{WL^3}{EI}$ at centre

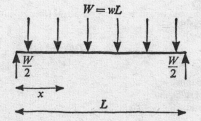

Figure 10.

For further information, see Roark, R. J., *Formulas for Stress and Strain* (McGraw-Hill, current edition).

Appendix 3   Torsion

Torsion

For a parallel bar or prism or tube under torsion the twist or angular deflection θ (in radians) is given by

$$\theta = \frac{TL}{KG}$$

where θ = angle of twist in radians

T = torque in inch-pounds or Newton-metres

L = length of member subject to torsion (inches or metres)

G = shear modulus (Chapter 12), N/m² or p.s.i.

K is a factor to be found from the following table.

Section	K	Max shear stress N
Solid cylinder		
radius r	$\frac{1}{2}\pi r^4$	$N = \dfrac{2T}{\pi r^3}$ (at surface)
Hollow tube		
radii r_1 and r_2	$\frac{1}{2}\pi(r_1{}^4 - r_2{}^4)$	$N = \dfrac{2Tr_1}{\pi(r_1{}^4 - r_2{}^4)}$ (at outer surface)
Hollow tube longitudinally slit (i.e. 'C' section)		
Wall thickness t Mean radius r	$\frac{2}{3}\pi r t^3$	$\dfrac{T(6\pi r - 1\cdot 8t)}{4\pi^2 r^2 t^2}$
Any continuous thin-walled tube of thickness t, perimeter U and enclosed area A	$\dfrac{4A^2 t}{U}$	$\dfrac{T}{2tA}$

Again, considerably more detailed information is to be found in Roark.

Appendix 4   The efficiency of columns and panels under compression loads

For a column

Assuming that the column is of such proportions that it is liable to fail by elastic buckling (Chapter 13), then the critical or Euler load P is given by

$$P = \pi^2 \frac{EI}{L^2}$$

where E = Young's modulus
        I = second moment of area of cross-section
        L = length.

Now suppose the column to have a cross-section which can be expanded or contracted while remaining geometrically similar so that its size is characterized by a dimension t, say.

Then $I = Ak^2 =$ constant $. t^4$
where A = area of cross-section
       k = radius of gyration (Appendix 2).

If there are n columns, the total load sustained

$$P = \frac{n\pi^2 EI}{L^2}$$

$$\text{so} \quad I = \frac{PL^2}{\pi^2 nE}$$

$$\text{so} \quad t^2 = \text{constant} \sqrt{\frac{PL^2}{\pi^2 nE}}$$

But the weight of n columns = constant $. nt^2 L\rho = W$, say where ρ is the density of the material.

$$\text{So} \quad W = \text{Const.} \, nL\rho \sqrt{\frac{PL^2}{\pi^2 nE}}$$

$$= \text{Const.} \sqrt{n}.L^2.\rho. \sqrt{\frac{P}{E}}$$

So efficiency of structure

$$= \frac{\text{Load carried}}{\text{Weight of structure}} = \frac{P}{W} = \text{Const.} \frac{1}{\sqrt{n}} \left(\frac{\sqrt{E}}{\rho}\right)\left(\frac{\sqrt{P}}{L^2}\right)$$

The parameter $\left(\dfrac{\sqrt{P}}{L^2}\right)$ is known as a 'structure loading coefficient' and depends solely upon the dimensions and loading of the structure. The parameter $\left(\dfrac{\sqrt{E}}{\rho}\right)$ is called a 'material efficiency criterion' and depends solely upon the physical characteristics of the material.

For flat panels

The above arguments apply to a column whose thickness can be varied in two dimensions. The thickness of a flat panel can only be varied in one dimension.

Suppose second moment of area per unit width of panel $= I = \text{Const.} \ t^3$

$$= \frac{PL^2}{\pi^2 nE} \quad \text{for } n \text{ panels}$$

$$\text{So} \quad t^3 = \frac{PL^2}{\pi^2 nE} \quad \text{Const.}$$

Weight of n panels per unit width $= W$,
$$W = nt\rho L \ \text{Const.}$$

$$= n\rho L.^3\sqrt{\frac{PL^2}{\pi^2 nE}}.\text{Const.}$$

$$= \text{Const.} \ n^{2/3}\left(\frac{\rho}{\sqrt[3]{E}}\right).L^{5/3}.\sqrt[3]{P}$$

So efficiency $= \dfrac{P}{W} =$ Const. $\dfrac{1}{n^{2/3}} \cdot \left(\dfrac{\sqrt[3]{E}}{\rho}\right)\left(\dfrac{P^{\frac{2}{3}}}{L^{\frac{5}{3}}}\right)$

Again $\left(\dfrac{P^{2/3}}{L^{5/3}}\right)$ is a 'structure loading coefficient'

and $\left(\dfrac{\sqrt[3]{E}}{\rho}\right)$ is a 'material efficiency criterion'.

Suggestions for further study

At the end of the day, the best way to learn about structures is through observation and practical experience: that is, by looking at structures with a seeing eye and by making them and breaking them. Of course the opportunities for the amateur to build real aeroplanes or bridges are likely to be rather limited; but do not be ashamed to play with Meccano, or even with old-fashioned building blocks. These things, incidentally, are much more instructive than the modern plastic toys which clip together in various ingenious ways. When you have built your bridge, load the thing up in a realistic way and see how it fails. You will probably be both surprised and disconcerted. When you have done this the rather dry books on structures will seem a good deal more relevant.

Although there is not much scope for the amateur bridge-builder, it has often seemed to me that the field is wide open in biomechanics. This is a new subject about which very little is known, either by the engineers or by the biologists. It is very possible that there is an opportunity here for the enterprising amateur to make a name for himself.

Though there are rather few good books, as yet, on bio-mechanics there are any number on materials and elasticity. A small and admittedly arbitrary selection is given below.

Books about materials

The Mechanical Properties of Matter, by Sir Alan Cottrell. John Wiley (current edition).

Metals in the Service of Man, by W. Alexander and A. Street. Penguin Books (current edition).

Engineering Metals and their Alloys, by C. H. Samans. Macmillan, New York, 1953.

Materials in Industry, by W. J. Patton. Prentice-Hall, 1968.

The Structure and Properties of Materials, Vol. 3 'Mechanical

Behavior', by H. W. Hayden, W. G. Moffatt, and J. Wulff. John Wiley, 1965.

Fibre-Reinforced Materials Technology, by N. J. Parratt. Van Nostrand, 1972.

Materials Science, by J. C. Anderson and K. D. Leaver. Nelson, 1969.

Elasticity and the theory of structures

Elements of the Mechanics of Materials (2nd edition), by G. A. Olsen, Prentice-Hall, 1966.

The Strength of Materials, by Peter Black. Pergamon Press, 1966.

History of the Strength of Materials, by S. P. Timoshenko. McGraw-Hill, 1953.

Philosophy of Structures, by E. Torroja (translated from the Spanish). University of California Press, 1962.

Structure, by H. Werner Rosenthal. Macmillan, 1972.

The Safety of Structures, by Sir Alfred Pugsley. Edward Arnold, 1966.

The Analysis of Engineering Structures, by A. J. S. Pippard and Sir John Baker. Edward Arnold (current edition).

Structural Concrete, by R. P. Johnson. McGraw-Hill, 1967.

Beams and Framed Structures, by Jacques Heyman. Pergamon Press, 1964.

Principles of Soil Mechanics, by R. F. Scott. Addison-Wesley, 1965.

The Steel Skeleton (2 vols.) by Sir John Baker, M. R. Horne, and J. Heyman. Cambridge University Press, 1960–65.

Biomechanics

On Growth and Form, by Sir D'Arcy Thompson (abridged edition). Cambridge University Press, 1961.

Biomechanics, by R. McNeil Alexander. Chapman and Hall, 1975.

Mechanical Design of Organisms, by S. A. Wainwright, W. D. Biggs, J. D. Currey and J. M. Gosline. Edward Arnold, 1976.

Archery

Longbow, by Robert Hardy. Patrick Stephens, 1976.

Building materials

Brickwork, by S. Smith. Macmillan, 1972.

A History of Building Materials, by Norman Davey. Phoenix House, 1961.

Materials of Construction, by R. C. Smith. McGraw-Hill, 1966.

Stone for Building, by H. O'Neill. Heinemann, 1965.

Commercial Timbers (3rd edition), by F. H. Titmuss. Technical Press, 1965.

Architecture

There are many hundreds of books on architecture. I have picked out two, almost at random:

An Outline of European Architecture, by Nikolaus Pevsner. Penguin Books (current edition).

The Appearance of Bridges (Ministry of Transport). H.M.S.O., 1964.

Index

FOR THE BEST IN PAPERBACKS, LOOK FOR THE 🐧

In every corner of the world, on every subject under the sun, Penguin represents quality and variety – the very best in publishing today.

For complete information about books available from Penguin – including Pelicans, Puffins, Peregrines and Penguin Classics – and how to order them, write to us at the appropriate address below. Please note that for copyright reasons the selection of books varies from country to country.

In the United Kingdom: Please write to *Dept E.P., Penguin Books Ltd, Harmondsworth, Middlesex, UB7 0DA*

If you have any difficulty in obtaining a title, please send your order with the correct money, plus ten per cent for postage and packaging, to *PO Box No 11, West Drayton, Middlesex*

In the United States: Please write to *Dept BA, Penguin, 299 Murray Hill Parkway, East Rutherford, New Jersey 07073*

In Canada: Please write to *Penguin Books Canada Ltd, 2801 John Street, Markham, Ontario L3R 1B4*

In Australia: Please write to the *Marketing Department, Penguin Books Australia Ltd, P.O. Box 257, Ringwood, Victoria 3134*

In New Zealand: Please write to the *Marketing Department, Penguin Books (NZ) Ltd, Private Bag, Takapuna, Auckland 9*

In India: Please write to *Penguin Overseas Ltd, 706 Eros Apartments, 56 Nehru Place, New Delhi, 110019*

In Holland: Please write to *Penguin Books Nederland B.V., Postbus 195, NL–1380AD Weesp, Netherlands*

In Germany: Please write to *Penguin Books Ltd, Friedrichstrasse 10–12, D–6000 Frankfurt Main 1, Federal Republic of Germany*

In Spain: Please write to *Longman Penguin España, Calle San Nicolas 15, E–28013 Madrid, Spain*

In France: Please write to *Penguin Books Ltd, 39 Rue de Montmorency, F-75003, Paris, France*

In Japan: Please write to *Longman Penguin Japan Co Ltd, Yamaguchi Building, 2–12–9 Kanda Jimbocho, Chiyoda-Ku, Tokyo 101, Japan*

FOR THE BEST IN PAPERBACKS, LOOK FOR THE 🐧

PENGUIN SCIENCE AND MATHEMATICS

Facts from Figures   M. J. Moroney

Starting from the very first principles of the laws of chance, this authoritative 'conducted tour of the statistician's workshop' provides an essential introduction to the major techniques and concepts used in statistics today.

God and the New Physics   Paul Davies

Can science, now come of age, offer a surer path to God than religion? This 'very interesting' (*New Scientist*) book suggests it can.

Descartes' Dream   Philip J. Davis and Reuben Hersh

All of us are 'drowning in digits' and depend constantly on mathematics for our high-tech lifestyle. But is so much mathematics really good for us? This major book takes a sharp look at the ethical issues raised by our computerized society.

The Blind Watchmaker   Richard Dawkins

'An enchantingly witty and persuasive neo-Darwinist attack on the anti-evolutionists, pleasurably intelligible to the scientifically illiterate' – Hermione Lee in the *Observer* Books of the Year

Microbes and Man   John Postgate

From mining to wine-making, microbes play a crucial role in human life. This clear, non-specialist book introduces us to microbes in all their astounding versatility – and to the latest and most exciting developments in microbiology and immunology.

Asimov's New Guide to Science   Isaac Asimov

A classic work brought up to date – far and away the best one-volume survey of all the physical and biological sciences.

The New Science of Strong Materials
or
Why You Don't Fall Through the Floor

Why isn't wood weaker than it is? Why isn't steel stronger? Why does glass sometimes shatter and sometimes bend like a spring? Why do ships break in half? What is a liquid . . . and is treacle one?

All these are questions about the nature of materials. All of them are vital to engineers, but also intrinsically fascinating as scientific problems. During the two hundred and fifty years up to the 1920s and 1930s they had largely been answered by seeing how materials behaved in practice. But materials continued to do things which they 'ought' not to have done. Only in the last forty years have these questions begun to be answered by a new approach.

Now, materials scientists, of whom Professor Gordon is one, have started to look more deeply into the make-up of materials – at the atoms and molecules on which their mechanical properties depend. They have found many surprises: above all, perhaps, that how a material behaves depends on how perfectly – or imperfectly – its atoms are arranged. If we could make a perfect crystal whisker of Epsom Salts, it would be far stronger than steel piano wire.

Now revised and updated – using both SI and imperial units – Professor Gordon's account of this fast developing science is a perfect demonstration of the sometimes curious and entertaining ways in which scientists isolate and solve problems.